DATE DUE

DEMCO 38-296

A SHEARWATER BOOK

PAUL SHEPARD

July 12, 1925–July 16, 1996

Traces of an Omnivore

McElway had been working quietly in one corner for more than an hour, measuring and sorting the fossil bits into piles. The daggerlike teeth and elongated tarsals told of centuries of carnage and flight. He put a short bone from some long-vanished foot and a rounded tooth out from the rest, saying to no one in particular, "These may go with that footprint we saw yesterday." The room fell silent, as though some invisible presence demanded our attention. "A planti-grade omnivore," he added, and for a moment we each had our private vision of something that had stood apart, even transcended all that pell-mell world of panicky prey and their voracious killers.

Carnassia Molar, *Trophism and Attention*

Traces of an Omnivore

Paul Shepard

Introduction by Jack Turner

ISLAND PRESS / Shearwater Books

Washington, D.C. / Covelo, California

Riverside Community College
Library
APR '97 4800 Magnolia Avenue
Riverside, California 92506

GF 49 .S54 1996

Shepard, Paul, 1925-

Traces of an omnivore

A Shearwater Book published by Island Press

Copyright © 1996 Paul Shepard

All rights reserved under International and Pan-American Copy-right Conventions. No part of this book may be reproduced in any form or by any means without permission in writing from the pub-lisher: Island Press, Suite 300, 1718 Connecticut Avenue NW, Washington, D.C. 20009.

Shearwater Books is a trademark of The Center for Resource Eco-nomics.

Library of Congress Cataloging-in-Publication Data

Shepard, Paul, 1925–
     Traces of an omnivore / Paul Shepard.
        p.   cm.
     Includes bibliographical references and index.
     ISBN 1-55963-431-6 (cloth)
     1. Human ecology.   2. Human evolution.   3. Deep ecology.
I. Title.
GF49.S54   1996
304.2—dc20   96-32772
                                                        CIP

Printed on recycled, acid-free paper ♻ ∞

Manufactured in the United States of America

10 9 8 7 6 5 4 3 2 1

Contents

Preface

I am told that there have been literary geniuses who foresaw their own futures and envisioned their careers with such clarity that they wrote with their collected works in mind. But for the rest of us scribblers there is no Destiny, only the issue at hand. Where we are headed we haven't the least idea.

There comes a watershed in the years when the future remains as obscure as ever (but less extended) and there is a kind of landscape of words behind us. We can see vaguely where we were, as though we had toiled up a mountain trail: so that's where we were going! It is even possible to discover or imagine patterns behind us, much the way meaningful figures can appear in leaves on the ground beneath trees in the fall, or, for that matter, in ink blots. As for the view behind, I can arrange my essays, written over forty years, so that they seem to have some transcendent purpose.

Still, there is some progression in them that is not imaginary. If so, it began for me with a certain book, a particular teacher, and the subject of eyes. The book was Gorden Walls's *The Vertebrate Eye;* the teacher was Rudolf Bennitt. For an hour or two a week, Professor Bennitt sat with me in his office while his daughter made us tea, and we discussed the chapters on eye shape and size, color sensitivity, the role of day and night vision, placement of the eyes in the heads of different species, anatomy and function of the parts—what seemed to me a marvelous demonstration of function as an aspect of habit and habitat.

In retrospect, the scene in my memory is not only of my own recollections but also of Walls's description of the backward looking as done by the rabbit without turning its head because it has bulging, lateral eyes, the better to judge the distance between itself and the jaws and talons of its pursuers.

Chasing me now is the demon-hunter of time and years. Having frontal, binocular eyes, I lack the rabbit's 360-degree field of view. My tracks, when I turn and look, have a fragmented visibility as the trail lengthens. In a longer perspective of time, you and I were not committed solely to the status of prey species. The demon years hound me, but I am not exclusively a leaf-eater. Like my fellow humans I am omnivorous and my vertebrate eye represents a long history of contemplating what is at hand, such as the reflections in a rabbit's eye.

Sight and perception are, in our evolutionary line, greatly influenced by vision—all the subtle influences that culture brings to understanding. Being neither a literary eagle who could see his destiny a mile away nor a rabbit-eyed scanner of the events taking shape behind me, I now see that the eye study was my first effort in consciously giving form to the past. Bennitt and Walls are part of that past, but so are a thousand generations of ancestors who were neither exclusively carnivore nor herbivore.

Walls and Bennitt were heretical in their time. In the mid-twentieth century it was widely believed that, however much the eyeball provided a sensory experience, vision was entirely different; that ideas had no habitat, being free of the apparatus of biology. That duality did not sit well; it was not what Walls was all about. I found myself—perhaps unconsciously—with an opposing belief about insight. What I saw in *The Vertebrate Eye* would later, in the third quarter of the twentieth century, be called "holistic." Eyes, brains, minds, cultures, and ideas were syncretic and continuous.

How we see events or issues always has a basis in the senses. Ecological reality is the arbiter of the senses and all other impulses in brains. It is the generator of mind and the sinew of culture. It is food habits and food webs, the passage of energy. I am ecologically an omnivore, and my attention is at least as versatile as my stomach and my location. Were I more keenly focused fore and rear, my spoor might be more obvious. As it is, these essays attest a record in twigs bent here and there, the odd muddy footprint, iotas of debris.

Having come out of Africa (only yesterday it seems), and from Professor Bennitt's office (was it this morning?), I have traveled West. It has been a trip into diversity. The signs of my personal presence are somewhat like those seen by the earliest emigrants on the Oregon Trail, for whom there was no "trail," only what they called "the trace."

Paul Shepard

Introduction
by Jack Turner

In his famous essay on Tolstoy, "The Hedgehog and the Fox," Isaiah Berlin distinguished between thinkers "who relate everything to a single central vision . . . a single universal organizing principle in terms of which all that they are and say has significance—and, on the other side, those who pursue many ends, often unrelated and even contradictory." The former are hedgehogs, the latter foxes, the basis for the metaphor being a fragment by the Greek poet Archilochus: "The fox knows many things, but the hedgehog knows one big thing."

Paul Shepard is a hedgehog. His central vision derives from two facts: First, despite ten thousand years of agriculture, several thousand years of urban living, and centuries of industry, technology, and science—despite, in general, all that we think of as progress—our genetic heritage, formed by three million years of hunting and gathering, remains essentially unchanged. As a species, human beings still inhabit the Pleistocene.

And second, if we accept the usual explanation of domesticity—controlled breeding to achieve a specific purpose—human beings are wild. At least for now. Genetic testing, and the de facto eugenics emerging from reproductive choice based on genetic testing, may eventually reduce our species to tall, lean, brilliant clones delighted with urban crowding, noise, high stress, tofu burgers, and virtual reality. But for the present we remain omnivorous mammals whose DNA is more closely tuned to the wilderness, "a world," as Shepard says in the essay "Advice from the Pleistocene," "to which our bodies and minds are already committed, a world essentially wild."

Shepard believes this genetic heritage influences everything from human neurology and ontogeny to our pathologies, social structure, myths, and cosmology. Regardless of one's view of the human condition, the belief that we are wild Pleistocene primates wandering around malls, playing with nuclear weapons, and systematically destroying our habitat is an original thesis about human nature. And that is, at bottom, what Shepard offers: a theory of human nature.

To a degree unusual in modern thinkers—most of whom have been cowed by the twentieth-century intellectual fashions of relativism, positivism, and postmodern deconstruction—Shepard answers the perennial question: Who are we? This alone would make him worth reading. But he goes further and mulls more difficult questions many people would prefer to ignore. What are the conditions under which human beings would not just survive, but flourish? Are we less healthy and happy in our domesticated landscapes than in the wilderness that was the ground of our being, the mold of our physical structure and psyche? Are the remaining hunter-gatherer cultures a benchmark by which to measure our decline? How is early contact with nature manifested in the creative functions of the mind—in metaphor, poetry, myth, metaphysics? Does the development of a healthy self-identity require, during childhood, bondings with animals, plants, a specific place? What happens to children who have no contact with Nature? How are we to become native to this land?

Shepard's answers to these questions—and to many more—are the subjects of the essays collected in *Traces of an Omnivore*. They have been culled from a variety of obscure journals; specialized, often out-of-print books; and the proceedings of various scientific organizations and conferences. Until now many have been difficult to locate. To have them all in one place is a delight.

The collection is also significant in that it contains material often buried or absent in Shepard's previous books. Especially useful, I think, are a series of five essays that honor his intellectual mentors. "Hunting for a Better Ecology" is an appreciation of Aldo Leopold's ideas and Shepard's most succinct statement of the importance of hunting and human ecology, an ecology "based on killing by hunting and gathering." "The Philosopher, the Naturalist, and the Agony of the Planet" explores Ortega y Gasset's *Meditations on Hunting*. After discussing Marshall McLuhan, Walter Ong, Erik Erikson, and the pernicious influence of Mediterranean cultures, "If You Care about Nature You Can't Go On Hating the Germans Like This" turns into an essay about Martin Heidegger. "Place and Human Development" acknowledges the influence of Edith Cobb and her landmark book *The Ecology of Imagination in Childhood.* And "On Animal Friends" elaborates and extends ideas associated with Claude Lévi-Strauss, especially in his famous work in *The Savage Mind.*

Perhaps of equal importance is another series—"Advice from the Pleistocene," "A Posthistoric Primitivism," and "The Wilderness Is Where My Genome Lives"—that contains Shepard's most recent ideas about our Pleistocene origins. These examples, however, only suggest the diversity of Shepard's interests. Among other things, the reader will discover discussions of landscape, aesthetics, the bear, hunting, perception, agriculture, human ontogeny, history, animals rights, domestication, postmodern deconstruction, tourism, vegetarianism, the iconography of animals, the Hudson River school of painters, human ecology, theoretical psychology, and metaphysics. Indeed, the diversity suggests a fox. But no—he is a hedgehog wearing a fox mask. Our wild Pleistocene genome remains central to everything he has written and the most conspicuous theme of this collection.

Shepard was the first person to hold a chair in human ecology. Although the term was in use as early as 1921, most of what had then been done in the field dealt with obscure issues in geography and demographics. In particular, no one had looked at the relations between the human mind, its habitat, and other species. It seems *Homo sapiens* had little interest in applying the ideas of ecology to itself.

But in 1973, Shepard, after teaching at Smith, Williams, and Dartmouth, was appointed the newly created Avery Professor of Natural Philosophy and Human Ecology at Pitzer College and the Claremont Graduate School. The new Avery Professor immediately set about applying ecology to human beings. By the end of his tenure he had put human ecology on the intellectual map.

Besides being impressive, Shepard's new title was uncommonly accurate. "Natural Philosophy" is an old category that until the eighteenth century included both natural science and philosophy. Shepard's thinking ranged across both. "Human Ecology," in contrast, was a new field, and, to use an adjective Shepard made famous, inherently "subversive." Together they suggested something like "subversive natural philosophy"—an accurate description of what Shepard was about to produce.

During his tenure at Claremont, Shepard played an important role in the creation of what are now called environmental philosophy, ecophilosophy, and deep ecology. He was a member of a brilliant group of thinkers, including the theologian John Cobb, Jr., and the political theorist John Rodman, who in 1974 created the Claremont Conference on "The Rights of Non-Human Nature." The participants were thinkers whose ideas helped define environmental philosophy for the next decade: Garrett Hardin, William Leiss, John Lilly, John Livingston, Joseph Meeker, Roderick Nash, Vine Deloria, Jr., Gary Snyder, and George Sessions. Each of them produced brilliant work, but none had Shepard's particular slant on what, after Earth Day, was described as the environmental crisis, and none would produce a body of work so systematically addressing human ecology—*Homo sapiens'* relation to its habitat and to other species. No one saw so clearly that the roots of the environmental crisis lay in the Pleistocene.

The importance of our Pleistocene origins was already present in Shepard's *Man in the Landscape: A Historic View of the Esthetics of Nature* (1967), a work connecting two themes that would dominate his future essays: contact with animals as a necessary ingredient of normal human development, and more broadly, contact with the natural world as a basis for both individual and environmental maturity.

Shepard then edited, with Daniel McKinley, two collections of readings—*The Subversive Science: Essays Toward an Ecology of Man* (1969) and *Environ/mental: Essays of the Planet as Home* (1971)—that introduced human ecology to a generation of students populating the new environmental studies programs that were springing up around the country. *The Subversive Science* became famous not only for its readings but for Shepard's essay "Ecology and Man—A Viewpoint," with its often quoted passages introducing his version of an extended self:

> Ecological thinking . . . requires a kind of vision across boundaries. The epidermis of the skin is ecologically like a pond surface or a forest soil, not a shell so much as a delicate interpenetration. It reveals the self ennobled and extended rather than threatened as part of the landscape and the ecosystem, because the beauty and complexity of nature are continuous with ourselves.

And further, in conclusion:

> If nature is not a prison and earth a shoddy way-station, we must
> find the faith and force to affirm its metabolism as our own—or
> rather, our own as part of it. To do so means nothing less than a
> shift in our whole frame of reference and our attitude toward life
> itself, a wider perception of the landscape as a creative, harmo-
> nious being where relationships of things are as real as the things.
> Without losing our sense of a great human destiny and without
> intellectual surrender, we must affirm that the world is a being, a
> part of our own body.

This was written before Earth Day, the Gaia Hypothesis, and Arne
Naess's historic article "The Shallow and the Deep, Long-Range Ecology
Movements: A Summary," announcing the arrival of deep ecology. It
remains as radical a statement as anything produced since then, a passage
that invokes science, poetic insight, metaphysics, and cosmology. Taken
seriously, it offers the prospect of a meritorious life with our planet.

When the self is expanded to encompass the world, environmental
destruction becomes self-destruction. Why would a species destroy itself?
During the decade beginning with his appointment at Claremont,
Shepard published three books that established his reputation. Each was a
blend of serious scholarship and informed speculation that demolished
ordinary academic categories, and each presented part of the answer to
this question.

The Tender Carnivore and the Sacred Game (1973), with its bizarre drawings by
Fons von Woerkom, remains one of the most penetrating studies ever
written about hunting and gathering cultures. *Thinking Animals: Animals and
the Development of Human Intelligence* (1978) examined the process of human
cognitive development and its relation to the diversity and complexity of
the natural world, especially animals. *Nature and Madness* (1982), which
Shepard believes is his most important book, combined the insights of the
previous works to outline a theory of human pathology. Shepard distinguished four historical periods—the advent of agriculture, the asceticism of
the desert fathers, the nature-hating of the Puritans, and the rise of mechanistic science—and suggested how each distorted normal human
ontogeny. The result, he claimed, was a civilization of "childish adults,"

several billion Pleistocene mammals living in a state of arrested development. Other authors—Erich Fromm for one—had claimed we live in a society no longer sane, but no one before Shepard had suggested that a stunted relationship to the natural world caused our insanity.

After this tour de force, Shepard turned to the study of animals and their importance in northern cultures, including our own. *The Sacred Paw: The Bear in Nature, Myth, and Literature* (1985), written with Barry Sanders, explicated Shepard's belief that the bear is the most significant animal for the history of metaphysics in the Northern Hemisphere. Ten years later he published *The Others: How Animals Make Us Human* (1995), a compendium of human-animal relationships that spans anthropology, zoology, evolutionary biology, philosophy, cosmology, and social criticism.

These books, together with many essays and innumerable lectures, constitute one of the most sustained attempts this century to understand our relation to the natural world. Yet despite his distinguished academic career, a prodigious list of publications, and a collection of awards, fellowships, and grants that would make most academics weep in envy, Shepard is not well known. There are several reasons for this. First, Shepard's books are formidably intellectual, devoid of nods to popularization. Second, many of his most important books are out of print.

Traces of an Omnivore helps correct this unfortunate situation. The essays discuss most of the major themes in his books. They are also lighter, more accessible—preliminary sketches of what would become demanding studies. If he were an artist, they would be his watercolors, not his oils. *Traces of an Omnivore* is thus a welcomed introduction to Shepard's ideas.

But there is, I think, a deeper reason for his relative obscurity: Shepard is a truly radical thinker. He may well be our most radical thinker about nature since Henry Thoreau.

To understand how radical Shepard is, consider this: the major environmental organizations could achieve all their goals and still not heal the pathology Shepard believes is destroying the Earth. Contemporary environmentalism is, of course, a complex movement with many agendas—all commendable. But this complexity is divided, broadly, by two views of how to mitigate our destructive relation to nature. Although we can, and do, support both, we should understand that each has a different focus.

One view, by far the most common, emphasizes what we can do for the Other. What can we do to preserve species and their habitats, to close the

ozone hole and cleanse the oceans, to slow—perhaps halt—the destruction of rain forests? This path leads to economic incentives, better
resource management, more and larger nature preserves, captive breeding
programs, pollution controls—the list is long. For each item on the list,
numerous scientific societies, political groups, and public policy organizations vie for public support. This is mainline environmentalism. It is not
radical because it doesn't offer what the term "radical" implies, both historically and linguistically: it doesn't address the root of the problem.

Since the root source of the Earth's destruction is human behavior, the
radical view requires us to assess who we are, why we are so destructive,
and what changes *in ourselves* might improve our relation to our habitat and
to other species. This path is often referred to as deep ecology. It is not
well understood and enjoys meager public support.

Furthermore, this radical perspective conceals a further schism. Most of
its advocates believe we can achieve in our present life an extended sense
of self, an identification with life in the broadest sense, whether described
as wild nature or Gaia, and that this identification will heal our rupture
from the natural world. Shepard is less sanguine. For him the salient issue
is human ontogeny, an optimum developmental schedule common to
every individual of the species; a wild, genetically programmed. Pleistocene ontogeny we must somehow regain. In short, Shepard offers an
explanation of who we are, why we are destructive, and how we might
change that emphasizes child-rearing practices, not choices we might
make as adults.

Shepard's claim is a scientific thesis. It is not concerned with individuals
of our species but with the species. Plant ecologists are not interested in a
pine next to your cabin—that is the province of the naturalist. They are
interested in *pines*. The same is true of human ecologists. But while there
are ecologies of bears, spotted owls, and whitebark pines, there is no
ecology of the human animal. For many people the very idea is presumptuous for the same reason evolutionary biology is presumptuous: many
people deny our species-hood, our relation to other species, past or present, and they refuse to consider the world of public and private property
as habitat. In other words, they deny human beings are part of nature.

Human ecology, like evolutionary theory and molecular biology, leads
inevitably to a theory of human nature, and any scientific view of human
nature must emphasize ontogeny. Most scientists agree that experiences in

early childhood interact with genetic endowment in ways independent of culture. Language acquisition, motor skills, and cognitive and emotional development are all keyed to specific time periods and require specific stimulation and feedback. If the feedback and stimulation are not present, or are present at the wrong time, the child's development will be permanently arrested. For example, children who do not hear a language or learn to mimic sounds will never master speech. Biological time slots hardwired into our genes during the Pleistocene place constraints on human potential. This is true if the individual in question is Japanese, Slav, Bushman, or Eskimo.

Some modern intellectuals do not believe in a "human nature" in the sense of something we all share, something intrinsic to our species. Looking at humanity they see, instead, a smorgasbord of cultures. However, a substantial body of evidence suggests that human life faces limits impervious to cultural diversity. As Shepard says in one of his most recent and interesting essays, "Wilderness Is Where My Genome Lives,"

> The paradox of an apparently unlimited adaptability and extreme specialization [in the brain and nervous system] will probably untangle its own contradictions in the twenty-first century, as we discover that cultural choices do not exhibit but hide common, underlying, physical limitations and requirements.

Assuming the genetic component of our ontogeny creates such limitations and requirements, there remains the issue of exactly how they are indexed to wild nature. Almost everyone who has written about human ontogeny has assumed that normal cognitive and emotional development results primarily from interactions with members of our own species. Similarly, virtually all theories of psychopathology have assumed that problems with these interactions cause character disorders—neuroses and psychoses. Shepard challenges both of these entrenched assumptions. His account of human ontogeny makes interactions with wild nature—animals, plants, and place—a necessary condition of normal human development. Deformations in these programmed Pleistocene interactions produce "ontogenetic crippling" and create pathological relations to our habitat, other species, and other members of our own species. Shepard's reasons for his position combine well-informed speculation with research on remaining hunter-gatherer cultures, especially the !Kung bushmen of the Kalahari. Like many people I know, I find them compelling.

Assuming Shepard's ideas in these essays accurately describe the roots of the environmental crisis, it is easy to see why most people would prefer mainline environmentalism with its pollution incentives and nature preserves. If the genetic basis of our minds, emotions, and self-identities must be triggered by temporally crucial contacts with plants, animals, and a specific place, and if, further, our destructive behavior as a species is a function of a "civilized" ontogeny that elides these contacts, then the mainline environmental agenda is pitifully inadequate.

But to move beyond mainline environmentalism will be difficult. Like Darwin's theory of evolution, Shepard's ecological view of human ontogeny displaces our privileged position in the universe. It underscores our radical dependence on precisely the wild nature we so readily destroy, it insists on seeing us as members of a species, and it requires us to make peace with our Pleistocene origins. This is unsettling, of course, but we need to understand *why* it is so unsettling.

Every ethical and political theory presumes a view of human nature. We usually ignore this theoretical baggage until someone comes along with a new theory that calls it into question, and popular consensus about human nature is quite different from Shepard's new ecological view.

For instance, Western cultures are underwritten by views of human nature that deny connection and dependence. They are, let us say, atomistic. We believe a "self" is a body harboring an ego at war with its id, clearly distinct from the surrounding world, and, if healthy, acting on that world autonomously. This is the heritage of Freud. We believe the basic social unit is an autonomous individual working out his or her self-interest under the watchful eye of the state. This is the heritage of the Enlightenment. We believe our species is separate from and superior to other species; we are, as we say, God's children, and the rest of creation is, and should be, subject to our whims. This is the heritage of Christianity.

Ecological thinking denies atomism in all its forms. Applied to human beings it promotes a radically different view of who we are. To act on it would require a revolution in human thought, and assuming we do wish to act upon it, there is the question of what we can do *now* despite the ontogenetic plight produced by ten thousand years of progress. Shepard must have been asked this question for most of his adult life. We cannot go back to the Pleistocene, right? So what to do? His rejoinders to this taunt produce some of the most enchanting ideas in these essays.

First, of course, we can't go back, because we have never left the Pleistocene. What we call history is merely one of the illusions that keeps us from seeing this fact. We are fundamentally the same. Our home, the Earth, is fundamentally the same. We have lost much, but much of what we have lost can be regained. The Pleistocene, in Shepard's words, is a "world vanished but not irretrievable." Changes in the scale of our lives, a new commitment to primary experience, the re-creation of vernacular styles and gender, a reacquaintance with wild animals, the acceptance of death, and, most important, alterations in how we raise our children—all this is possible. The task ahead, however littered with controversy, is to recapture forms of social life appropriate to our biological heritage.

Paul Shepard is familiar with controversies. The essays in *Traces of an Omnivore* address them with an intellectual courage uncommon in an age that exults the relativist, the skeptic, and the cynic. Perused with care, they will reward the reader with a deepened appreciation of what we so casually denigrate as primitive life—the only life we have in the only world we will ever know.

Plants, Animals, and Place

The Ark of the Mind

There is a profound, inescapable need for animals that is in all people everywhere, an urgent requirement for which no substitute exists. This need is no vague, romantic, or intangible yearning, no simple sop to our loneliness or nostalgia for Paradise. As hard and unavoidable as the compounds of our inner chemistry, it is universal but poorly recognized. It is grounded in the way that animals are used in the growth and development of the human person, in those priceless qualities which we lump together as "mind." Animals have a critical role in the shaping of personal identity and social consciousness. Among the first inhabitants of the mind's eye, they are basic to the development of speech and thought. Later they play a key role in the passage to adulthood. Because of their participation in each stage of the growth of consciousness, they are indispensable to our becoming human in the fullest sense.

In the first twelve years of every child's life, animals are seen by the imagination directly, without interpretation. Unencumbered by symbolism, they are as plain and unambiguous as their names: horse, cow, dog, chicken, bird, elephant. Each is for the child part of a fauna of behavioral conventions: whinnying eagerness, bovine nurturing, yapping pursuit, clucking anxiety, aerial capering. In its particular way of behaving, each calls up in the child a latent feeling or idea. The characteristic "message" of each animal is thus an outer reference whose corresponding inner twin is brought to life in the observing, mimicking, inchoate human polymorph.

This matching game is the first game for individual consciousness. Impressions received by the child from the intense activity of pretending, staring, and naming are forms of actual nourishment, providing sustenance for the eventual synthesis of the self and the growth of symbolic thought. Beast by beast, in the first years of life, the emotions, feelings, attitudes, intentions, and fears take their place in the forest of the self. Only through this process is the child free of the ambiguity of abstract explanation. The example of other humans is not a sufficient training ground for the perceptions. The adult's quicksilver change of mood, shading of trait, blending of response, tempering and concealment of motive, and fluid shifting of inner shape to fit outer circumstance are too slippery. Human subtlety is important later; but for the child it can be a form of madness.

A decade, from the beginnings of speech to the onset of puberty, is all we have to load the ark. The zoology of this period must be unequivocal, without recondite allusions. Poetry and song must mean what they say, games must be nothing but play, as unmistakable as a cat chasing a ball. It is right for the child to mimic fox and goose in a game of pretended capture, or speak the lines of the little pig or Chicken Little. By "identifying" with a number of animals in turn, the child discovers a common ground with other beings despite external differences between himself and them. Anthropomorphism at this stage is essential. The true means of inter-species communication, full of invisible nuance and removed from sensory detection, is not yet pertinent to the tasks of the child. By pretending that animals speak to one another, he imposes on them a pseudohumanity which, although illusory, is the glue of real kinship.

In such farces of socialized ecology, the vital natures of animals are encountered—and become our best defense against the conspiracy that animals are only machines or artifacts, and therefore against the lie that we ourselves are made of cogs, wheels, and wires. It is important as well for the child to literally see the animals' insides, for organs have names too, thus forming a fauna of stomachs, lungs, and hearts to which our own belong. Only the child who has had this experience can be pleased by his or her own organic nature.

Much is at stake in the first decade, for it culminates in a bonding to the matrix of the earth, a crucial step between the first infant-mother bond and formal entry into adult membership in a cosmos. Here the foundation for a poetry of ultimate meaning is based. This matrix, in which animals are the living, animate aspect, will be only as ample as the child's observation of nature, linked to speech and mimicry in play. The metaphysical rich-ness of the individual's eventual personal philosophy depends on it. During childhood, however, abstractions such as chemistry, physics, ecosystems, morals, and ethics are noxious.

The animals do not live in an arbitrary environment (except for zoo and barnyard). Like their bodies and behavior, their location is peculiar to each. For human beings, habitat and environment are the literal space of the ground of thought. As messages, animals come into thought trailing the dust of their associations with a particular place. What the child wants, as Edith Cobb observed, is to find a place in which to make a world the way the world is made. The home range of the ten-year-old is the first context of spatial and temporal thought, perceived unconsciously in harmonious replication of his mother's body—the first "place" in contact with which the fetus and newborn moved. The child is a "traveler" mapping out the first spatially ordered reality of his life. The habitat of childhood is conceptualized as an ordered space inhabited by its creatures—turtles, frogs, mice, and rabbits—as events in place. For a ten-year-old, the home terrain is thus a constant pattern in which the compelling actions of animals are like moments in the life of a great spatial being.

The end of childhood is the end of that simple identity. The literal fauna have become the external expression of the child's own congeries of feelings and bodily processes, a community of self-confidence. That confidence will soon be tested, for adult life is full of contradictions. Indeed, adolescence is a preparation for ambiguity, a realm of penumbral shadows. Its language includes a widening sensitivity to pun and poetry. Appropriate to its psychology is attention to the zones between categories, zones that have their own animals. The borders from which obscenity and taboo arise are figured in creatures that embody a sense of overlapping reality: the insects that crawl between two surfaces, the owl flying at dusk, the bat that seems to be both bird and mammal.

The adolescent person is a marginal being between stages of life, on the shifting sands of an uncertain identity. In this respect his symbols are changeling species: the self-renewing, skin-shedding snake, the amphibious frog that loses a tail and grows legs, the caterpillar that metamorphoses into a butterfly. In each the thought of a new birth is manifest, the concrete expression of transformation. Human psychogenesis is such that the adolescent is, for a time, plunged back into his own natality. The concreteness of life, literal in the maternal and natural matrices, given consciousness in speech itself, will be reexperienced in a new, metaphorical idiom. No echo of this infantile state is more crucial than nourishment. Eating, the most fundamental route from outer to inner, is to be reevoked as the ritual act at the core of transformation and relatedness. Its emotion is refocused in intellectual and symbolic ways, using incorporation as the metaphor of connectedness. Henceforth, all rites of passage elevation in social status, marriage, the reception of spiritual life—are celebrated and sometimes represented as feasts. Sacred meals, taboo food, and dietary laws everywhere refer to what is eaten as an agent of change in the eater.

As a collective, the animals of the natural environment comprise the metaphor of the human group. In tribal culture, each clan is committed to a particular species. This species, through its ecological relationship to other species, provides a vehicle both for the dynamic logic of myth and for the rules of society. Together the clans constitute the whole in a manner analogous to the ecology of animals. For the relationship between clans is defined by the relationship between their totemic animals in accordance with a myth about the animals in the beginning of time. Both this myth and the observations of the creatures themselves guide the interrelationships of humans, who are pledged by their clan identity to the mythic structure. The same fauna mimicked in childhood play to synthesize the individual self is, in maturity, liberated into new levels of social and metaphysical deliberation.

The use of animals in play in the first decade of life gives way in the young adult to dance, a universal human activity derived from the rhythmic imitation of animals. Through dance, in traditional societies, a particular human group acquires a style of its own, uniting its members while at the same time affirming the tutorial role of birds and mammals. People have always suspected that certain animals are masters and keepers of important secrets: metamorphosis, birth, puberty, healing, courtship, fertility, and protection. By dancing the animal, these mysteries are assimilated into adult understanding and recovered as a power of humankind.

Part of becoming adult is the dawning realization that the principle of transformation is a major feature of the cosmos. Movement and passage-making are inseparable from consciousness of time. Dancing in the feathers of birds and the masks of mammals displays the shape-shifting capacities of the soul. The religious principle of altered states has its special animals, whose greatest in the Northern Hemisphere is the bear. From its natural history comes a rainbow of horological suggestion so powerful that it may have changed the history and evolution of human thought. In circumpolar traditions across America, Europe, and Asia, south to the Gulf of Mexico, the Mediterranean, and the Himalayas, for perhaps fifty millennia, festivals of the bear ceremony have recognized the bear as sacred messenger and mediator, purveyor of meat, the paradigmatic grandparent, teacher, traveler between worlds.

Many features of the bear—especially the many races of the brown bear—place it in correspondence to humanity. Its size, appearance, mobility, dexterity, omnivorousness, reproduction, annual cycle, length of life, social behavior, and intelligence have an eerie relation to our own. These characteristics are the source of enduring speculative analogy and psychological tension. The geography of this rapture is as wide as the distribution of the brown bear and as ancient as mankind, a whole paragraph in the zoological hieroglyphics of human consciousness.

The bear is the only familiar omnivore whose size approximates our own. Omnivorousness is not only a kind of diet but a versatile style of perception—exploratory, pushy, relentless, zetetic, analytical, risk taking. It is like tasting the fruits of all actions, the meat of all situations, the kernel of all experiences, the root of all being. The bear is fisher, hunter, berry-picker, bulb-digger, honey-gatherer. It has an expressive face, binocular vision, vocal and gestural responses, sitting and bipedal stances, almost no tail, and a fine dexterity. Mother bears give birth secretly, tend and teach their young, and defend them fearlessly. And yet the bear is vividly Other— huge, furry, long-muzzled, long-clawed, quadrupedal—in these things nothing like a human.

In winter the brown bear withdraws into the earth from which it came as a cub, this winter sleep coinciding with the death of nature. Spring comes after the bear's emergence, as though it were called by him back to life. In the perspective of years this seasonal passage into the earth becomes a rhythmic movement. The bear's trip into the earth, translated into the rhythm of the life cycle, is unmistakably about death and rebirth.

In the hunt for the bear, there is no chase; it is killed ceremonially. A terrible, dangerous animal when abroad, in its den the bear is easily slain. Such a hunt of a large mammal is unique, for the animal is located as though it were a plant, seeming almost to combine the hunt with gathering. The meat of this "stepmother" is ritually distributed by the men, but the women dance and sing—reuniting what cannot otherwise be healed. The bear is given. More conspicuously than any other animal eaten by us, bear flesh is a gift to man from a distant god, showing that all hunted game allow themselves to be taken.

Spiraling in the northern sky, the celestial bear drops below the horizon in step with the seasonal sleep. In the world underground the bear dwells with its own people. To come into middle earth and provide us with sacred meat he puts on his bearskin and is welcomed by the mistress of the

hearth, who shares the bear's life-giving secret. The people who understand these things butcher him with reverence and eat him carefully, confirming his special reality in themselves. In part of the message of the
hunted bear is:

> Save the bones of your dead and inter them in the earth.
> Remember that the spirit survives and lives again. Connect this
> sacred quality with every individual in a ceremonial bear-meat
> feast of communion. There are parallel lives below and above
> your plane which are eternal. Passage between them is the ulti-
> mate movement by which you know life.

The bear is the keeper of all gates: those between life and death, this
world and others, flesh and spirit, man and animal form, inside and outside,
even the phases of human individual life. He is the mediator between man
and woman, the natural and the sacred. All guides and travelers to the
other world in human form—shamans, Orpheus, Jesus—were represented
by bears first.

* * *

The warping of the animal out of the myth of resurrection is a historical
development, the collapse of an instructive metaphor. Replacing the bear
exclusively with the human figure denies us each our bearskin. By zealously
repudiating the animal form, omitting the middle matrix, we retreat from the
polymorphic ambiguity of life. The bearless cosmos deprives us of personal experience of the sacred paradigm, substituting for it abstract, verbal
exegesis. The loss makes for autism, middlemen, desperation, the failure of
the kindred species who think in us.

The carrying of a positivistic, literal attitude toward animals into the
adult sphere marks the failure of initiation and maturity in human life. The
totally humanized myth of immortality is part of the zeitgeist of domestication, its ritual centered on sacrifice rather than the sacred hunt. Our
dreams, however, remain true to a world different from the one in which
we now live. Hunger for the wild animal's significance is reflected palely in
the vicarious imagery of decorative arts, virtuoso and eccentric originality,
pets, and media stereotypes.

No fine words can replace the dances and feasts of participation. Those
arts remind us that we were thought up by the different beasts. They are

kindred and ancestors. Before humans existed they worked out the round of life in thousands of variations as though anticipating the needs of style in the experiment of human cultures. Like the bear, we are selves composed of sleeping figures, each a secret that can be awakened in acts of correspondence. Self-consciousness is possible only in a world of Others. We are members of a human family and society, but the presence of animal Others enlarges our perception of the self beyond the city to the limits of the world, and deeply inward to that ground of being where live the lizard and monkey and fish.

Animal Rights and Human Rites

A progressive march of civil justice over the past two centuries is the historic setting within which blacks, women, and lately eighteen-year-olds have been in part elevated in political power and social status. Whole nations and economic classes composing a "third world," usually identified by a nonindustrial economy, are now popularly regarded as unjustly deprived. The implication seems to be that powerful nations are economically obligated to satisfy the right to be rich and, implicitly, to be free. Meanwhile the egalitarian flood reached out to orphans, working children, infants, idiots, and cripples. Then, with the humanitarian movement of the past century spilling over, it came to the protection of domestic animals—a kind of fourth world—invoking in their name the right not to be overworked, tortured, or abandoned.

The movement for the rights of animals and plants would seem, therefore, to be an extension of the legal means of civilization. The question is now whether the furthering of an ethical relationship to animals can indeed come as an *extension* of the idea of human rights. Is a civil and legal foundation derived from several millennia of purely human politics a sound basis for ecological justice? If not, what is the context of such an ethic?

Today's concern for the well-being of animals in zoos and medical laboratories is a further unfolding of the defense of homeless dogs and cats. It is typical of this protective attitude toward animals to be yoked to the theme of poor housing: the condition of caged animals, and finding homes for strays. There is something about the "homelessness" of wild animals that obsesses humanitarians. For them a bird's nest is obviously its home, and even a hole in the ground is better than no roof over one's head. This fixation on shelter plays into the hands of such enterprises as the logging and paper industries, whose reseeding of the plains of their disasters is publicly described as "creating a home" for wildlife. Like antilitter and beautification themes, such propagandist lullabies are the tunes to which protective associations find themselves swaying, if not dancing.

It also has an Oriental expression, a logic by which rights of creatures, so frustrating in their semantic and practical aspects, can be serenely refined to the purity of nonkilling. A recent exponent of this view was Albert Schweitzer, who, combining St. Augustine with Hindu Jainism, was devoted to the protection from death or injury of all life, however minute or inconspicuous.[1] Schweitzer's "reverence for life" was an endless chain of excruciating decisions in all things medical and horticultural. His unshakable conviction of the rights of life over death and of the priority of human value did not mitigate his somber sense of judgment and responsibility. For him, too, the care of life was associated with giving shelter. He kept captive a small menagerie of wild animals so that he could admire them. As among all wild captives made into pets, these animals were amputated from their gene pool and from fellow creatures and habitat; no amount of loving care could prevent them from becoming neurotic and flabby monsters.

Schweitzer's impulse to "extend to all life that respect which we have for our own," however much influenced by Eastern thought, is clearly related to Western humanitarian social justice. It was a part of the colonial mind, a kind of obligation to those "less fortunate" on a scale where rank is confirmed as much by responsibility as privilege. Charity has always had something of the readiness to be resigned to a status quo. That the relationship between nonhuman species should require human intervention, though as old as the "dominion" precepts in Genesis, could only be acted upon by those who knew themselves to be the instruments of God's will. Even so, Schweitzer is no exception to the general sense of alienation that preoccupied European philosophy and literature at the time, for it was a strategy in which humanism and technology combined to prove human transcendence.

Like all ideologies, this attitude could perceive ecological conflict as an uncompromising *us* or *them*. The consequences are either exile or sanctuary. As any naturalist knows, there is a widespread general assumption that any wild creature that is no longer found here has simply gone somewhere else. The popular discovery of the migration of birds had the unfortunate effect of seeming to support the notion of a kind of infinite mobility. It took half a century to convince duck hunters that the ducks no longer flying south no longer existed.

The concept of sanctuary is more interesting, for it is no mere illusion. It is the only humanitarian "solution" available to the ideological imagination. The American Indian and the American bison thus went onto reservations at the same time. The idea has the merit that it recognizes the multiplicity of factors necessary for life and treats survival as populations instead of individuals. Continued existence demands a space, geography, or habitat, which cannot be fully described and is not known. We now have whooping crane sanctuaries, condor sanctuaries, black-footed ferret sanctuaries. In New Zealand I have seen a frog sanctuary, all the known individuals of which live on a single rocky ridge no bigger than a football field.

Most such protected space is "set aside" for relic populations of certain endangered species. Reservations are a kind of reprieve for those creatures whose range and number have so shrunk that they are in danger of becoming extinct, a last-minute commutation from a death sentence to a life term. Or, in a political idiom, we may imagine all survivors of such minorities transplanted to a new state. Sanctuaries are also likened to museums established by patronizing largesse for the edification and amusement of visitors.

The example of the sanctuary for frogs points up the inevitable problem in a world of two million species. Scores of species of molluscs and freshwater fish are either endangered or already extinct. The TVA alone accounted for several species of snails. No one knows of these—or cares—except a few naturalists. It is obviously impossible to establish a sanctuary for every creature which must soon face that ultimate ideological moment. Such will not be necessary. The humanitarian jury has an ancient Chain of Being etched in its conscience. "Worthy" species can thus be chosen, and it is not likely that we will go much "below" frogs.

If the theme of sanctuary were actually followed to its logical end, with every endangered species in turn granted its own reservation, the spaces would gradually coalesce until the whole planet was a sanctuary. Not a bad idea, perhaps. But this cannot happen, because the concept is a political solution to nineteenth-century problems when space was still unlimited. In evolutionary terms this unfeasible arrangement could be described as allopatric—the simultaneous life of potentially conflicting species by geographic separation. Extinction, exile (as extirpation), and sanctuary are the allopatric choices in humankind's cohabitation of the earth with other life.

Allopatry is consistent with our traditions of personal property, the model of the nation–state, the philosophy of the domination of nature, and even with industrial growthmania up to a point. The alternative is sympatry, or living together. One form is captivity for the nonhuman, which is an indirect form of its destruction as a species. We are nearly without the cultural rudiments of an ecological ethic based on sympatry. Only the Peaceable Kingdom offers an image of that halcyon state. Unfortunately, it refers to no creatures that we know. Although depicted with the outlines of lions and lambs, the inhabitants of that mythical garden have no other resemblance to life.

There are some unchosen sympatric symbiotic forms. The dooryard birds, weeds of the lawn, roaches, fleas, house mice—a few dozen stow-aways in the wheelhouses and holds of spaceship earth. It is a sparse fauna and flora, to which the pests and diseases of mankind may be added as species in their own right. Though wild in the true genetic sense in which domesticates are not, they are relatively unimportant to the larger question. That both vegetarian and other ethical nonkilling philosophies are exercised primarily on these urban fellow travelers and domesticated animals is indicative of the irrelevance of the creatures and the practices. Extinction has almost nothing to do with direct killing (with some exceptions, such as whales). The preoccupation with not killing houseflies is like a narcotic. Whole realms of life are swept away while devotees bicker over the sanitation of zoos and zealots gingerly move each ant from the sidewalk. It is almost as though some devil had provided this homely assembly of houseflies and milk goats to divert us.

Our society has hardly begun to address the question of sharing the land with competitors, predators, parasites, and with those forms so little threatening and so fragile that special care is needed to prevent their inadvertent destruction. The scientific evidence, good as it is, cannot guide us to economic decisions affecting all daily contacts with a million species or with the environments on which they depend. The situation is illustrated by the self-serving argument of chemical pesticide industries against restrictions: "Get all the evidence before you make a decision"— a provision that would postpone regulation indefinitely. For the same reason—that we do not know all the details of all the niches of all crea-tures—the concept of rights (at least as conventionally conceived) cannot be applied. We do not in fact know what must be done to assure the most elementary right of survival of more than a few hundred species.

Something much more pragmatic than economics or quasi-political rights is called for: procedures that keep our connectedness with all life before us, even though we don't know how the connections work. Indeed, certain universal elements of ecological systems might be taken as representing the whole. I have in mind food chains. In contrast to the ethical-humanitarian canon of no killing, the food chain, or trophic-ecological, approach would be to emphasize the centrality of predation. The humanitarian's projection onto nature of illegal murder and the rights of civilized people to safety not only misses the point but is exactly contrary to fundamental ecological reality: the structure of nature is a sequence of killings. Part of the "moral" difficultly of accepting this criterion is that it is overlaid with the overwhelming fact of the murder of some seventy million people between 1913 and 1946. The notion of deliberately linking such a dangerous "blood lust" to an active philosophy of nature must surely on its surface seem monstrous. There will be no alternative but to begin at the beginning, to examine the nature of war as part of the politics of the members of a single species and the nature of predation as a wholly different matter.

The Cynegetic Sensibility

Have humans ever lived by a code of interspecies killing who do not kill one another? If so—and I believe that the evidence is strong for such peoples, both in our past and among living hunter-gatherers—what can it signify for us? Admittedly, at this point it can do no more than to signal a hopeful surface. We must unravel the so-called "primitive" one step at a time, peeling back the alternate layers of anthropological fact and cultural lies, hoping that illumination from a time and a way of life to which we cannot literally return will shine upon us.

Not only are war and murder rare, but among those killers of animals, whose lives were a perpetual celebration of food chains, there was and is disinclination to maximize. The jaw of the economist in us falls open; here were and are people who did not take more when there was more for the taking. No wonder the nineteenth century regarded the "savage" as having a "lack of drive," "deficient intelligence," or "physical apathy." It was as incomprehensible as the notion that such low matters as killing and eating can be close to the center of any kind of ethical system.

Apprehension of connectedness to nature is precisely the weakness of Western ethics. Marriage and diet are the universal means and symbols of relationship. It is just in the matter of food chains in nature that the cynegetic (hunting-gathering) mind excels. For us, animals in general constitute a vague Other; we are reminded throughout our lives of the many distinctions between us and them. It is extremely difficult to recast that whole assumption, to break out of our own culture by an act of will, to see the Others as a myriad of beings, the differences among which are on a plane with the differences between us and any one kind of them. In the cynegetic view the world is an exquisitely elaborated comity laced through with eating habits, the most conspicuous and important threads of connection and significance. When one senses, however momentarily, that experience of a world of beings which it produces, the loneliness of modern humankind can be seen as an ecological as much as a social phenomenon. No one could be lonely in a world so richly populated.

For hunters, that tapestry is a hieroglyph in which the patterns of eater and eaten are the principal figures. The terms of this communication are species, and the grammar their habits. The extensive knowledge among hunting peoples the world over has astonished ethnologists for generations. That the Fang of the Gabon, the Hanunoo of the Philippines, the San of the Kalahari, and dozens of other peoples have plant and animal taxonomies numbering in the hundreds and even thousands was merely one of those irrelevant (or merely "practical") but irritatingly ubiquitous facts until this knowledge was related to (1) the individual development of cognition and (2) the mosaic reference for totemic metaphor by way of an accompanying natural history. "The diversity of species furnishes man with the most intuitive picture of his disposal and constitutes the most direct manifestation he can perceive of the ultimate discontinuity of reality," comments Lévi-Strauss.[2] But the infinite discontinuity is not a fragmentation, because of the natural history; the myriad forms are connected by food chains, learned as the nomenclature is learned. Only continuity as patterns of a flow relationship saves the necessary refinement of perception from becoming a nightmare of pulverized Otherness.

That we academics today scorn the memorizing of names reflects our culture's contempt for nature. As Victor Shelford and other pioneers of American ecology discovered in the 1920s, the study of the dynamic life of natural communities had to wait on inventory and recognition. Even among the sciences taxonomy is a poor relative. Every high-school boy interested in wildlife recognizes electronic apparatus and fast vehicles as the symbols of truly chic research.

The use of the food chain as an instrument in the perception of creature-rights in sympatric existence grows from the evolution of human thought. Such is the evidence of anthropology. But to bring the food chain into the sphere of every human experience as the language of relatedness requires a more explicit and personal poetry, the precisely honed metaphors of ritual ceremony. Joseph Campbell suggests that the only effective rituals are those that refer to something homologous in each individual infancy.[3] In this case the primordial association is with oral nutrition. In suckling, we mammals prepare a vision of the world.

The point is that to describe food chains verbally is ineffective; one must use a deeper language. That ritual ceremony has deeper roots than speech is evident in its universality. Its very use by us is a sign of continuity with nature. Among vertebrate animals the important information is communicated in stereotyped, unequivocal signs derived by "stylization" from some basic care-giving behavior. Thus is a drinking posture ritualized by gulls to affirm a pair-bond or the exaggerated yawn of the baboon as a threat. Such signals are always communicative, species-specific, given in context, innate, and often in tandem or cybernetic arrangements. They range from momentary displays of bright feathers to extended human religious ceremonies, a continuum that seems objectionable only to those committed to doctrines of separatism. The whole range of ritual, from courtship in birds to Renaissance art, was surveyed under the aegis of the Royal Society of London in 1956. Since then, ethological work, studies of human behavior, and the psychological study of art—particularly Franco-Cantabrian cave painting—support the unifying thrust of that symposium under Julian Huxley's leadership.[4]

Although we can never know what ceremonies were held in the Paleolithic caves of southern Europe twenty thousand years ago, there is a substantial argument that the game mammals portrayed were perceived in a metaphor of human dimorphism, and that the painted forms on the naturally rounded surfaces functioned as transitional objects in the psychiatric sense of forms intermediate between self and not-self. They are formulations of trophic status transformed by the philosophy of a great Magdalenian culture.

Only beings that have a place in a coherent creation can have rights. Ultimate questions of personal and species identity turn not only on the question of what one eats or is eaten by but upon the silent wisdom that unity is achieved by the specific eating of kinds and parts, a merger of natures and mutual assimilation thrown into relief by that which is not eaten and by the anastomosing of the one food chain with the many. If the indigent state of taxonomy shows the low regard of the insular human intellectual community for the gnosis of kind and category, the secular meal must be its public counterpart.

Might saying grace have ecological consequences? Indeed, it may be impossible to overestimate the achievement of interspecies ethics that would follow a renewal of ceremony at table. In a single generation there might come into existence a society imbued with the joyful solemnity of assimilation as children and weaned on its sacred meaning in adolescence. The implications would carry beyond the particular organisms consumed, since the homologizing nature of ritual would spread sanctity across the whole of life.

This would fly directly in the face of the humanitarian impulse. Instead of turning from death—and therefore from life—dietary ritual explicates the rights of animals in the why and by whom they are killed and eaten. In its absence, consider the middle-class meal, in which children are normally unreflective or conditioned to be silent regarding the death on the plate before them and who, when reminded of it, react out of repugnance for all things bloody and organic, by which a fastidious society has trained them to deny the sensuous spirituality of living tissue, or they indulge in the sentimental tantrums that prevent the human heart from childhood to senility from truly loving the Other. Only from the right to die and be eaten properly, in the proper and acknowledged food chain, can the right to life for all creatures become apparent.

Albert Schweitzer's "reverence for life" seems an appropriate term for this. But for him there was no room for predation at all. He remarked that it was a perpetual sorrow that life feeds on life. He could not break out of a two-thousand-year tradition in which the loss of paradise was associated with the daimonic metaphor of the apple of paradise—the coupling of eating and sexuality, which is a central image of venatic thought. It is not surprising that two thousand years of this denial of life produced civilizations in which all eating is regarded as merely bodily and sensual instead of sacred, guilt is a perpetual side dish, and every animal stands prejudged for all time.

Of course this paints an incomplete picture of our society. Communion is a Christian rite of assimilation by eating and drinking. If ecologically it seems like holy cannibalism, it at least retains the potential forms for an interspecies ceremony. Mealtime grace is still widely observed. Even if it thanks an abstract provider instead of the creatures eaten, it at least calls attention to a primal analogy and keeps alive nutritional mysteries.

Gratitude for the meal is only a beginning. Before it comes the cooking, butchering or preparation, killing or finding. Among the living peoples for whom the sacred and profane are not sealed off from each other, every stage in the food quest carries on its back some special observations and protocol. It remains to be discovered whether children have critical period requirements in these activities. Is witness to butchering-before-eating, for example, an irreplaceable event in the life of a five-year-old by which the articulation of one's own insides is achieved? If so, butchering is an "educational" event in the organization of internal reality and perhaps even an archetype of a larger reality.

As any coroner, surgeon, or anatomist knows, what is revealed in dissection is not simply a standard suit of entrail underwear, like the parts of a certain kind of automobile. What can be learned is never complete, as every individual is a departure in detail from a basic schema. Besides its candidacy for rites of assimilation, every corpse is an omen. Its anatomical eccentricities are surprises that signify on several levels. A liver is a message about the quality of food and drink antecedent in the food chains and therefore the quality of the soil that supports them. In addition to these causal connections, the organs flesh out possible shifts in the social entities to which they are analogous and seem to bear oracular power. Every animal to be eaten is first a message that we ignore at our peril, its language garbled where animals are domesticated, and denied to modern urban children.

The words "sacred," "sanctity," and "spirituality" have been used in these remarks related to the trophic apparatus of human ecology. The poetry of the food chain and its mythic potentiality are better perceived in religious than legal terms. The rights of men and women in society can be articulated and are understood to be social and political creations, even though they are thought to incarnate something intrinsic. But the rights of nonhuman forms are their behavior. Since there is no alternative except death, "right" is synonymous with their continued existence. When nonhuman rights are dealt with directly in the courts, a farcical parody of justice occurs—as in the fourteenth century, when roosters and donkeys were burned at the stake or given other punishment. Certainly humans can be constrained by legal means in their actions toward the nonhuman. The difficulty is that "continued existence" refers to species, not to killing versus nonkilling. Like Aldo Leopold's definition of conservation as "what a man thinks while chopping" a tree, it is less a question of *what* is done than of *how* it is done.[5]

Zoomorphizing in the Quest for Identity

Apart from food chains, there is another universal interpenetration with nonhuman life that could be evoked in the name of the whole. The emptiness of the modern human psyche is twofold: at one end the loss of identity in food chains and at the other a more immediate failure in the perception of self. If only we knew all the other creatures perfectly, it would leave exactly the cognitive space for humans; we would know who we are as humans without having once looked into a mirror. We are endlessly engaged in search of surcease from our inchoateness. Throughout life we put ourselves in a kind of cosmic police lineup, now beside an oak, then beside a dolphin, then a chimpanzee. Of each we can say, "We are not this . . . and yet, there is just this that we share." With each discovery we place one piece in a puzzle.

Of course the process is not that conscious or deliberate. Instead we "take up the other" as a metaphoric strategy in play and language, in a drama of predication, of the suspension of disbelief while assuming another identity to find our own. Before we can be subjects to ourselves we must become objects. To think as humans we must first think as nonhumans. To think nonhumans is to enact them. Does this zoomorphizing seem merely to compound the error of anthropomorphizing, already so widely regretted by intellectuals?

Both may have a place, but they are not coequals. Humans are zoomorphs, while other creatures are not humans. Zoomorphizing places us in the context that protects us from that anthropomorphism which empties the taxonomic pyramid like a funnel and remakes the diverse forms of life in our image. Zoomorphizing connects us to the plurality of Otherness.

The imitation of animals by children in games of leapfrog, the investment of personality in a stuffed animal toy, the animal stories and animated films, the animal-headed gods, the metaphors in language ("He is a skunk!") are more than illustrations or even analogies. By incorporating into one body the mix of self and not-self, the wholeness of diversity is confirmed in the moment before we sort them out. The body—the organism—is the archetype of connectedness. In this way the process of locating ourselves bit by bit in different natures does not shatter us like glass.

The metaphoric recapitulation has its own ontogeny, a progressive unfolding in each human life. For the seven-year-old playing fox-and-geese it functions differently than among adolescents playing on a football team called the "panthers," and differently still in the caricaturing by cartoons on an editorial page. A lifetime of nonhuman sign images stretches from them to the troubled subjectivity of the human self and group, arching through each life more or less ceremonially.

It may be asked at this point whether this widespread use of animal metaphors has anything to do with the rights of animals. Is this not just another example of people's self-centered exploitation of their fellow creatures? True, we "use" animals and plants as metaphoric fillers in successive frames of reference, making them into instruments of thought and perception. But in the process they also become themselves. The effect of locating them in a cognitive framework flows both ways: self-recognition and niche recognition. After all, the same accusation of exploitation can be directed to the eating of animals, which I have already discussed. The question itself is biased by ideology. The ideological question represents the no-compromise idealism of dichotomy-choice. The biological form of that imagination is the two categories: human and animal. In that view, if rights of the nonhuman are simply for the nonhuman to be what they are, the amorphous nature of nonhuman amounts to no right at all.

Another question may be raised as to whether the taking of animal identity by metaphoric predication actually requires natural species. James Fernandez speaks of this "primordial" predication in childhood as a stage from which a maturing ritual capacity progressively refines its transformation terms.[6] He cites the ceremony of the Eucharist as an adult example. Nowhere does he explicitly say that the human capacity for liturgical rites requires a childhood fixation on animals, only that animals are conveniently ubiquitous. Even if one were to acknowledge the necessity of animals at some critical period in triggering the child's poetic sense, why would imaginary animals not do just as well?

Moreover, the predication is not that of an animal itself but of a sign-image. To momentarily take the point of view of the bull or the wolf does not require the presence of an actual bull or wolf. If the stuffed toy is a vehicle by which the culture transmits these sign-images to the child, why not a soft, furry "wook" or "gleefus"—not a creature from familiar nature, but a new invention for which no natural prototype is necessary and whose natural history can be equally fictitious?

The question goes to the heart of the matter. It opens questions about the relation of art to nature that cannot be gone into here. It bears, moreover, on the revolution in perception of the world by which domestication and agriculture separated themselves off from hunting-gathering, and, as Lévi-Strauss says, the model of reality was shifted from its ancient place in nature to the world of made things as totemism was replaced by caste. Still further, the question bears on the place of monsters in psychology and culture.

In a study of the sculptured animals on Gothic cathedrals Walter Abell made the extraordinary observation that there was a correlation over some three hundred years of French history between the amount of social stress and the amount of stylization of the carved figures.[7] He concluded that periods of tension and unrest, times of famine, plague, or war, coincided with abstract forms, and alternatively that social and ecological harmony was signified in naturalistic forms. Much the same thing can be seen in nineteenth-century landscape painting. Presumably, we cannot accept a world whose purposes are inimical to our own. In good times nature can be taken at its face value, but in bad we must look to its fundamental structure below the visible surface for the necessary support that its externals deny.

Monsters and imaginary forms are to this extent mind-creatures of desperation, emergency sign-images. In a sense no imaginary creature is possible: if we give the "gleefus" legs or a head it has some reference to the zoological realm. The virtual impossibility of imaginary beings, unfettered by natural features, suggests that the unthinkable is unthinkable because of the necessity of creature prototypes or metaphors in the genesis of human thought. Did the human mind invent the ecological niche system, or is mental activity an expression of the species system?

But the question remains—any number of objects in our environment may satisfy the resolution by metaphoric predication that relieves one more iota of our poor pronoun nebulousness. Can we not say that some of us are more like Fords and others like Oldsmobiles with some profit?

No doubt such objects can serve the function, but the problem they raise has to do with content. A quick survey of the stuffed creatures in toy shops confirms an unconscious and unarticulated consumer demand— that the definition between kinds be kept, that identity remains important. The distinctions go far beyond the leopard's spots, extending to its behavior as well. These features are not invented; they are empirical. The ultimate reference is the real animal. The silent demand by children for keeping that reference sharp is evident in the naturalism of the toys. There is a demand taxonomy implicit that is also associated with the hierarchic structure of taxonomic categories—species, genera, families, and so on. Made and fictitious objects suffer from an inevitable ambiguity of systematic relationship as well as behavior.

Some of the inchoateness of our lives can never be resolved, though we are destined to continue trying. "What does it mean to be alive?" is one of those elemental enigmas. No one outgrows such primary questions. We may leave the paradigm of the animal to small children only at the risk of subscribing to the doctrine of the great Chain of Being, wherein the natural is lower or simpler. To the biologist, those who call machines "alive" are toying with deep semantics. The predication of the Ford or Oldsmobile, the taking of machine shape and the transient identification with it is a dangerous game to play. This is not because it threatens human dignity—an injury to our pride—but because it displaces us from a primary task of marking out a place in a cosmos that is given and which is more terribly Other than anything we can make.

Lévi-Strauss is concerned with the intellectual use of natural things for the purpose of social categorization—the formal metaphoric predications of totemism. It is but a part of the perennial need to distinguish between humans and other creatures, the preoccupation in human thought with the separation of culture and nature. I mention this here to remind us that what has been said on the preceding pages is not intended to support the pop-ecological theme of man's *identity* with nature. Pretending to be an animal is to obtain a momentary identity for the purpose of partly breaking it. In this the two formalities—the food chain and the predication of sign-image—have certain things in common.

They both are the obtaining of a whole being to unite with it, to reject the "inedible" while assimilating certain of its qualities. Such creatures are not arbitrarily obtained by the individual; the individual's culture has selected the subsystems of trophic and paradigmatic biota. Both refer to natural history and extend into the environment from mind and body through connections that are discerned in nature: to hunt is to know the habits of the hunted; to enact is to know the habits of the model.

Metaphor

The quest for the nature of the human self is linked to an information pool found in the biosphere. The information content of an ecosystem is related to its genetic complexity, organized into species; it is transmitted primarily by reproduction and predation. The transformation of this information by the human quest is largely through metaphor. This activity discerns not only separate species in their uniqueness but the centrality of the food chain. Thus, for example, food exchange in nature is the archetype of exogamy—the information flow through marrying-out that relates clans one to another. The eating—the food chain—is the language of mythical heroes and animals.

Communications regarding status therefore involve ritual aspects of ingestion. In any specific ceremony it is right that certain things be eaten as much as that certain ones eat.

Ecological study has firmly established that the well-being of prey species is as much at stake as that of predators in food chain systems. The "struggle" of the individual prey to elude the predator has to do with sorting out particular sets of genes for continuation, not for the escape of the prey as a species. By eating certain individuals and not others the predator becomes the prey's instrument for filtering the information to be transmitted to its own succeeding generations. Moreover, the hawk's "choice" of eating a rabbit instead of a mouse may be looked upon inversely as an option within the ecosystem whereby part of its total information is in that instance transmitted along certain lines and not others. In these examples it cannot be asserted that the hawk "uses" the mouse or rabbit any more than the mouse uses the hawk or the ecosystem uses them both.

The human use of animals as food for thought may be looked upon in the same light. After several hundreds thousands of years of the cultural exercise of metaphor, who has chosen whom? Like the options in the food web not exercised on any given occasion, can we say that the natural things for which *this* culture has no poetic use are unimportant? Every exercise of these predications furthers a relationship. How arbitrary can a conception slowly built up in this way be? One test of validity is in the consequences in human action. Whole ecosystems may live intact or be impaired. Therefore every species in them has some stake in the outcome. If the basic acts of metaphoric predication are instinctive and unconscious, then people are no more aware of what is accomplished than the animals they caricature. Can we say that they "choose" the elephant to imitate in a certain ritual any more than that the elephant has chosen to direct some of its impact on the environment via this route instead of, say, pulling up acacia trees? Even if we grant that the elephant is indeed an arbitrary cultural choice, we cannot so easily claim that the necessity of making *some* choice is an optional device created out of thin air, as that is inconsistent with the evidence. The very plurality of increments in this lifetime process requires an equally complex field of reference. Indeed, the complexity of relationships that are perceived demands many more kinds of creatures, habitats, terrains, and places than are incorporated in these enactments.

If, as I have suggested, the continuity of these processes results in both the human and nonhuman members having "vested interests," then what emerges is a field of relationships not unlike that of parts of the human body. Is the question "What are the rights of my foot?" meaningful? Only to the extent that it calls attention to contiguity. Suppose we could see only the foot and did not know that a human being dangled from it? Would we establish laws concerning the protection of feet? If so, what about the thousand other parts of the body similarly perceived?

It is a poor analogy perhaps, but it will have to serve the conclusion. What we do instead is to undertake a formal reticence, a restraint resulting in its protection. For every species we behave as though there were an unseen human attached to it—some part of our humanity in jeopardy.

But this perspective does not really evoke social action or legal policy. Since the connections are unseen they are best referred to indirectly. The official legislation of poetry, drama, or religion is generally unsuccessful. They are their own best arguments. To begin some slight shift in our general sense of acknowledgment or affirmation may be all that is possible—a rite no more revolutionary than a few moments' reflection on the consequences of buying prepared food: a society wolfing down meat it was too chicken to kill and prepare, activities going against the grain because they soil one's hands.

NOTES

1. See Albert Schweitzer, *Philosophy of Civilization*, vol. 2, *Civilization and Ethics* (London: Black, 1923); *Out of My Life and Thought* (New York: Holt, 1933). Also *The Animal World of Albert Schweitzer*, ed. and tr. C. R. Joy (Boston: Beacon Press, 1951).

2. Claude Lévi-Strauss, *The Savage Mind* (Chicago: University of Chicago Press, 1966), 137.

3. Joseph Campbell, *The Masks of God*, vol. 1, *Primitive Mythology* (New York: Viking, 1959).

4. *Philosophical Transactions of the Royal Society of London*, B-251, 1956.

5. Aldo Leopold, *A Sand County Almanac* (New York: Oxford University Press, 1948), 68.

6. James Fernandez, "The Mission of Metaphor in Expressive Culture," *Current Anthropology* 15(2): 119–146 (1974).

7. Walter Abell, *The Collective Dream in Art* (Cambridge: Harvard University Press, 1957), ch. 14.

Phyto-resonance of the True Self

Selfhood is a combined conscious and unconscious construction, aided by the capacity to refer intangible aspects of one's being to conceptual images borrowed from the outside world. In general, only characteristics of the sensible world are represented mentally. Clinical and cultural evidence suggests that a coherent concept of the inner world of the self is compiled by reference, by imagining evanescent aspects of one's being (organic function, emotions, and social relations) as well as the existential reality of the viscera. The environment constitutes a repertoire of connotation, not as a casual reference, but as an essential part of the evolution of cognition. Awareness and the manipulation and communication of abstruse reality is achieved by linking it to specific, external configurations—especially visible forms—which can also be reproduced in art.

Animals constitute a major class of connotation, to which certain inner events are keyed. This metaphoric device serves the individual in speech, in mythic and poetic thought, in therapeutic meditation, and in dreams. One application of this resonance is between the image of the animal and of the experience of some physical or psychological peripatetic quality. For example, Eligio Gallegos has been extremely successful with the imaging of animals in the therapist-client setting, in which animals associated in the patient's imagination with the chakras (the body centers of spirit, thought, voice, heart, action, emotion, and base) are invited into low-intensity "conversation" and therefore serve as voices for concerns otherwise buried. The procedure implies an active role for the imaged animals and their corresponding affinity for the feeling and thinking functions traditionally associated with the six energy centers. Such visualization does not require the physical presence of real animals in the meeting. Its emphasis is on performance, in which the animal corresponds to events that move us.

I suggest that plants function in a similar fashion and that together they represent a little-known but widely experienced holographic correspondence between the natural world and the mind. The

27

analogous plant-human encounter might have different characteristics. A phyto-resonance—the reciprocation of an internal aspect of the self and an external plant—could act at more fundamental levels than that of animals.

Compared to affective states, what biologists call "vegetative functions" (digestion, assimilation, growth, circulation, metabolism, and so on) are at once more intimate to us as basic activities, and yet more elusive in the difficulty we have in ordinary experience of perceiving them as part of our being. They do not lend themselves metaphorically to the active voice of animal surrogates. Lacking the humanlike features of animals, plants and plant communities present themselves as externalized elements of the self that are less assertive. Our rootedness in the earth and the spatial qualities of our relationships based on place are imprinted unconsciously, available to a botanically sensitive internal organizer, a resonance to which we are intrinsically predisposed and psychologically committed by our ontogeny.

The history of mythology is rich in signs of plants' affinity for evoking aspects of one's inner life. One need only remember the radiant mandala effect of the rose, the lotus, or the cross section of a tree trunk; King Solomon's "garden enclosed"; the visionary, gemlike, preternatural luminescence of flowers as doorways to "another world"; the syllogisms of symbolic fruits and seeds; and the trees of life and of the knowledge of good and evil. We are inclined by modern culture to see these references as literary or artistic devices. This huge body of symbolic allusion has been tainted by a thousand years of the logic of the parable and homily, or, in the modern world of virtuoso performance, by the assumption that natural appearances are merely the raw matrix of creative analogies in art. Poetic references to rootedness, flowering, and fruitfulness are misunderstood as arbitrary rather than essential processes. As figures of speech, literary similes miss the point.

While there are individual plant "presences" and parts to which we respond with allusive insight, such as "stock" and "scion," there may also be a distinctive function of the whole plant association. Freud, for example, speaks of the unconscious as a forest wilderness. This reminds us that plants, more often than animals, are perceived as a collective, from the leaves on a tree or grasses in a field to trees in a forest. In this they seem to connote tissue rather than organs, the organic contexture rather than affect. The landscape appears as the equivalent of the body itself. Its living forms correspond to internal organizations, from cellular to organ systems, to their different activities, to "feeling functions" and emotions. Even the terrain, with its fine-grained, dispersed, continuous mineral characteristics, may resonate in our consciousness to elemental aspects of the self, to skeletal or even molecular structure.

The unique quality of these metaphors is that plants keep their places and constitute holding ground in nature and thought. There are radial, sessile, marine, and animal forms, and there are plant life stages that move or are moved, but the ordinary distinction that animals are mobile and plants are stationary holds perceptually. The similarity of blood and sap notwithstanding, the exceptions to such generalizations do not vitiate our normal definitions, so that the elements of the self to which the floral world is metaphor tend to be the embedded character of organs and tissues.

Because of the vegetative structure of the external world we spontaneously perceive ourselves as densely organized, diverse, living, patterned as components, arranged in three or four dimensional patterns, self-healing, home to other, smaller life that may hide in and move through our tissues, green and innocent to ourselves, brittle and brown in age. It may be that the more useful plant metaphors for the less assertive aspects of the self are the analogy with involuntary functions or between habitat and the organ system or plant kingdom as the spatial dimensions and sustainer of the living system.

I have noticed with interest that in preparing his participants for meditation on the animal "speakers" of the six chakras, Eligio Gallegos engages his clients in envisioning themselves as germinating seeds reaching for the sun and have wondered if this botanical preliminary was not based on establishing the best facilitating environment for the "dialogues" that follow. There is a prior reality to the existence of animals, even of the mind, an herbaceous substrate. Moreover, animals are commonly seen as playing out a role, while plants are silent hosts, yet more powerful in their root growth than any burrowing animal. Although they represent a storied structure of animal spirits, totem poles are wooden, after all, providing the medium of contiguous substance and holding ground for figures. Wood also represents the original growth that made them possible for the carver's knife, and ensures their durability.

In studying the history of gardens, I found it helpful to think of them as conceptualized miniatures of the cosmos. In this vein, the Garden of

Eden is the world before it was stained with death, the thirteenth-century maze is the landscape of lifelong quest and perpetual risk of error and deception, in which the Right Path is scaled down to a ritual or a game, and the seventeenth-century gardens of La Notre or Boyceau are the universe of God the geometrist, reduced to small worlds. As gardens, plants function tangibly in these collective images in ways to which animals are not suited.

I now find the alternative to be equally interesting, if not true, that the garden (like the forest or the prairie) is something made large rather than contracted. It is the inner landscape projected. The great outside may therefore be read as it amplifies not only intangible qualities of the self but sets of those inner relationships that are more structural than the boisterous life of desire, anger, and anxiety, which flit through our being like animals. As we move through mazes, formal gardens, or gentleman's parks we are engaged in ways of thinking that reveal our internalized, cultural notions of faith, logic, and intuition. In gardens, created unconsciously for this purpose, we discover our own mysterious insides, perhaps in different styles at different times to meet the contemporary needs of the self-image. In the small, informal garden we deal with plants more individually or in some more intimate, vernacular sense of the inner world of the private self.

At the risk of offending anyone whose aching back and dirty hands testify to hard work in the garden, I suggest that as either cosmic structure condensed, or internal terrain writ large, gardens are a spatial and organic metaphor that deals with being rather than doing. Collapsed or projected, the garden metaphor mitigates against the modern impression that cognition, like the body, is essentially androidal—an idea reinforced by the inanimate structures, electronic and mechanical devices, with which we surround ourselves.

Carl Pribham and Joseph Chilton Pearce suggested some years ago the idea that the human brain is functionally a hologram of the universe in the same sense that the body is a metaphrase of a niche in the ecosystem. This idea is consistent with the observation that the human brain is layered, the older or inmost at the bottom, like sedimentary rocks. Different parts of the brain and nervous system correspond to aspects of the external world. The mind is implicated in these layers of the brain, as its neural structures represent the evolution of the brains of ancestral mammals and vertebrates. These cognitive centers have available, or are predisposed to acquire, certain external correlates of their own ineffable processes. With the evolution of human cognition, aspects of the outer world became internalized as a code for envisioning or experiencing otherwise inconceivable or alienated aspects of the self. Presumably one may exercise the code in visualization, as in Gallegos's work with animal imagery, but surely the ultimate implication and application, especially in our encounter with plants, is the palpable role of organisms in nature.

What can all this mean in how we understand plants in our lives? (1) It is based on the assumption that the perception of a flower, seed, plant, garden, or prairie spontaneously refers us to fugitive aspects of the self, a reference of which we are normally unaware; (2) that encounter has physiological (or psychophysiological) consequences, especially in terms of healing; (3) these interspecies interactions between ourselves and plants are ecological as well as psychological, having their origins in the co-evolution of our species and plant associations; (4) the beneficial aspect of living in a plant-rich environment may have ontogenetic parameters, as implied in the concept of the *kindergarten*, with its overriding concern for the child's development; (5) there may exist circumstances of deprivation as well as therapeutic applications that can best be addressed in the art form of the garden; (6) the correspondences between the plant and the human individual have yet to be catalogued; and (7) what the environment provides, in addition to concrete images, is coherence that not only makes the self available to the self but gives it wholeness.

I have used the idea of the person-plant metaphor in the sense that metaphor is defined by Elizabeth Sewell as essentially organic and tacit. If the human self is indeed dispersed perceptually in the landscape, to be discovered there incrementally, a whole new meaning is implied in the phrase "hunting and gathering." While anthropologists no longer think of these activities as strictly gender-bound, the plant-female and the animal-male associations in the long Pleistocene preamble speak now to the differences in body-perception of men and women. Now, as our culture assumes that the organic environment is simply a neutral substance for material exploitation, we must inevitably fail—to paraphrase Edith Cobb—to make a world in which to find a self the way the self was made.

NOTES

Janet Bord, *Mazes and Labyrinths of the World* (New York: E. P. Dutton, 1975).
Edith Cobb, *The Ecology of Imagination in Childhood* (New York: Thames and Hudson, 1988).

Roger Cook, *The Tree of Life, Image for the Cosmos* (New York: Thames and Hudson, 1988).

Eligio Stephen Gallegos, *The Personal Totem Pole: Animal Imagery, the Chakras, and Psychotherapy* (Santa Fe: Moon Bear Press, 1982).

Aldous Huxley, *The Doors of Perception* (New York: Harper, 1954).

John Mitchell, *The Earth Spirit, Its Ways, Shrines, and Mysteries* (New York: Crossroads, 1975).

Harold Rollin Patch, *The Other World According to Descriptions in Medieval Literature* (Cambridge: Harvard University Press, 1950).

Joseph Chilton Pearce, *The Magical Child* (New York: Dutton, 1977).

Elizabeth Sewell, *The Orphic Voice* (New Haven: Yale University Press, 1960).

Bears and People

Like us, the bear stands upright on the whole foot, eyes nearly in a frontal plane, so that we look into a true face. He moves his forelimbs freely in their shoulder sockets, is nearly tailless, sits up like a child or slouches like an adolescent, worries, moans, sighs, courts with demonstrable affection, snores, spanks the offspring, loves sweets, and has a distinct moody, gruff, or morose side.

He is a creature in his own right, needing no justifying human likeness, but we cannot shake off the impression that behind the long muzzle and beneath the furry coat is a self not so different from ourselves. By habit, the bear is a kind of ideogram of ourselves in the wilderness, as though telling of what we were and what we have lost. He is wily, strong, fast, agile, and independent in ways that we left behind when we took to agriculture. In moving away from his presence, we also surrendered a worldview that held him in reverent awe. He symbolized the harmony of society and nature, disrupted in the civilized world in a philosophical lurch whose pain we still feel, separating us from our natural origins.

Even so, our long association with bears is imprinted in daily language, religious concepts, folklore, fairy tales, place names, toys, plant and food names, surnames, and secretly in a hundred words that jointly stem from old Indo-European concepts: barrow, bier, barn, burden, bring, bereave . . . The bear lives in our urban hive as a teddy, a family in "Goldilocks," Smokey, or just as an athletic team.

In the figure of the bear we sense a secret power, perhaps the embodiment of our anima, a knowing or kenning. Bear myths around the Northern Hemisphere tackle fundamental questions of human existence. The stories are a paragraph in the idea that nature is a language. There are other sacred animals in the traditions—raven, coyote, tiger, elk, eagle, kite, whale—but the bear has a role in the mind for which it is endowed beyond them all.

Its power is framed in its natural history, most poignantly in the onset of winter as green things seem to perish and the soil freezes. Except for conifer needles, the leaves wither and fall, the sounds of frogs and insects cease, and the days grow cold and short. Many of the birds and small mammals vanish. With snowfall, the pale face of death comes over the world. The caribou and elk dig for moss or dried grass, pursued in turn by the wolf and other predators. Men too endure, hunt, and savor the berries and nuts put by. In time past the animals were both food and teachers of food-getting, scavenging, storing, and surviving. The bear especially spoke to the season's inescapable analogy to the life of the individual human. Perhaps over many centuries the human question went beyond "How do we survive the cold winter?" to "How do we survive the cold death?" And the bear had an answer.

It was, of course, enacted rather than revealed, but was no less astonishing. The bear's passage into the earth, under the burden of winter, buried beneath the snow like a punctuation in the round of life. The bear's miracle was double, for the female emerged with young, conflating birth and rebirth. The bear comes out just ahead of the snowmelt, as though its own heat set the new year in motion, timed to graze lightly on the tendrils of the first sedges, to gnaw the carcasses of frozen reindeer, bighorn, and deer revealed by the shrinking snow mantle.

As a kind of master of the wheel of the seasons, of the knowledge of when to die and when to be reborn, the bear heads the list of animals who vanish from the land in winter, a phenomenal retreat and rejuvenation timed to the greater pulse of coming and going. In its den, despite its ascetic existence, without eating or excreting, the bear "knows" the solar cycle, and the she-bear's meticulous motherhood is a private miracle.

In its tutorial defeat of death the bear's gifts to humankind did not end. It was itself a dedicated immolation to the human belly. Taken from its crypt in the snow and killed for a midwinter festival, its viscera would heal a dozen winter complaints; its fat was like liquid gold. People had a feast of meat to warm their insides and a hide to wear or sleep in. Rare enough not to become ordinary food, big enough to delight in a time of need, the bear-slaying was the obverse side of the bear's death. Like the sleep of death in the den, it too would only be a stage, and the hunting peoples saw themselves as instruments of its immortality. Summoned by such a gift, the humans were invited to intrude in the middle of the bear's deathlike sleep, to receive its body as nourishment and achieve spiritual awareness by participating in its reincarnation. From feast to festival, stomach to sacrament, the sacred celebration of the death of the holy bear became the great rite of the northern world.

Like all such ceremonies, it referred to an explanatory, exemplary, wonderful tale. Its details vary in different parts of the world, but the Haida version from British Columbia is typical:

Long ago, a group of girls of the tribe were gathering huckle-berries. One among them, a well-born young woman, was a bit of a chatterbox, who should have been singing to tell the bears of her inoffensive presence instead of laughing and talking. The bears, hearing her even though some distance away, wondered if she was mocking them. By the time the berrypickers started home the bears were watching.

As she followed at the end of the group, the maiden's foot slipped in some bear shit and her forehead strap, which held the pack filled with berries to her back, broke. She let out an angry cry, using the prohibited word, "Bear!" The others went on as she stopped to pick up her fruit, complaining. It was growing dark.

Near her appeared two young men who looked like brothers. One said, "We will help you with your berries. Then come with us, as it is late." As the aristocratic young lady followed them she noticed that they wore bear robes. It was dark when they arrived at a large cavern near a rock slide high on the mountain slope. All of those inside, sitting around a small fire, had put their bearskins aside.

Grandmother mouse quietly scurried up to the girl and squeaked to her that she had been taken into the bear den and was to become one of them. Already, the hair on her own robe was getting longer. She was frightened. One of the young bears, the son of a chief, came up to her and said, "You will live if you become my wife."

As the wife of the young bear chief, she tended the fire in the dark house. In the winter she was pregnant, and her husband took her to a cliff cave near the old home, where she gave birth to twins, which were half human and half bear.

One day her brothers came searching for her, and the bear husband knew that he must die. Before he was killed he taught her and the Bear Sons the songs that the hunters must use over his dead body and the other things they must do to ensure their luck. He willed his skin to her father, who was a tribal chief. The young men then smoked him out of the den and speared him. They spared the two children, taking them with the Bear Wife or Bear Mother back to her people.

The Bear Sons removed their fur coats and became great hunters, guiding their kinsmen to dens in the mountains and instructing the people in singing the ritual songs whenever a bear was killed, sharing their festival with its spirit, thanking it for the gift of meat and skin, and ceremonially sending it home to the mountain, pleased with its visit, so that it could return in another year.

In this Bear Mother story the abductors of the maiden, obviously alarmed about the immorality of human blasphemy, assume human form. The woman is mated with a divinity, becoming herself a sacred procreator, in a form of hierogamy. The tellers acknowledge kinship with an ancestral bear, yet keep the distinction between themselves and bears. The holy bear dies for the welfare of people, exacting atonement and propitiation. He is the spirit of all bears and will judge the moral state of human hunters thereafter, as he occupies the bodies of bears. Immolated, as food, slain bears become sacramental objects.

The Bear Sons, themselves in part divine, share in both mundane and spiritual attributes. As hunters they are intermediaries who teach that hunting is a sacred activity and understand the religious procedures of reconciliation between the hunters and the hunted. All humans are descended from the Bear Mother, the founder of the clan. The story explains why all "luck" in hunting is the result of an exacting ceremonial acknowledgment of the divine gift of meat, hide, and medicine, even when their immediate sources are other animals and plants.

The Bear Mother story may be the first great mythopoetic account of all life—the first external incarnation of our personal mothers—before written history and deeply affirmed in the human psyche.

Such important stories have a way of being written in the stars and of being reflected in a parallel universe beneath the earth. This vertically divided the cosmos, the sky becoming the home of eternal beings, the earth surface as the domain of mortal creatures and humans, and the underworld as the place of the dead, before their ascent or renewal. A corresponding horizontal partitioning of the world grew from bear mythology. This is the river of life and its people—its headwaters the home of immortal souls, its middle span that of the living, its lower reaches the netherworld.

An offshoot of the story, more popular in the world of adventure and brave deeds, traces from the Bear Sons, perhaps the prototypes of all heroes and their quests and surprising reappearances. Classical philosophers would concur that the idea of *thereutes* or *venator* embraces all quests for booty, love, understanding, and wisdom, linking them more closely to a masculine model of society and deemphasizing the mother.

The three layers of the cosmic universe have an odd association with forms of the infinitive *to bear.* Thus did the verb *bear* become a major part of speech. All the meanings of *to bear* congeal in three: *to bear* as in navigating, *to bear* as in carrying something, and *to bear* as in giving birth. These terms elicit the bear's exemplary actions in the three realms—sky, earth, and underworld. The location and movement of the stars is the basic guide to location and travel. The middle earth of daily life is where we encounter the bear messenger as a deity. And the netherworld is the essential place of transformation and renewal in the endless round of life and death.

In the Sky

Reference to the celestial bear may seem strange for such a great, lumbering animal. But what we have now turned over to astronomy and meteorology was once a major human concern and preoccupation: making sense of the night sky. An approach to the problem must have been reached early in human prehistory, since all of humankind now shares it. That is, the upper world is where the important stories are illustrated. This central idea developed some extremely esoteric expressions, but all of them rely on the notion of a zodiac, and the older they are, the more animals there are in them.

The Great Bear (degraded in recent centuries to a pictorial image, the dipper), Ursa Major, and the Little Bear, Ursa Minor, shine brightly in the heavens, among the most conspicuous constellations of the Northern Hemisphere. The brightest star in Ursa Minor, Alpha Ursae Minoris, the Pole or North Star, is the extension of the earth's axis. Both constellations are composed of seven stars.

The Hindus, who probably got it from Caucasian invaders, call the seven stars of Ursa Major the Seven Bears or *rishis* (wise men), and the Sanskrit name was *rakh* (bright). "Rakhtos" was likely confused by the ancient Greeks with their own word for bear, *arktos*, resulting in the bear's turning up in the sky. But there are more interesting possibilities, also preserved in Hindu lore.

For example, they believed that the Great Bear was the source and wellspring of all the energy in the universe. As one writer, De Gubernatis, says in *Zoological Mythology*, it caused the "seasons to follow one another in regular succession, rains to fall, and crops to grow and ripen. . . . It assured . . . a supply of food, but if it gave . . . health and strength, it also, as the controller of water and wind, caused droughts in season, and sent blights and diseases on evil winds."

The association of bears and birds is not only their mutual connection to the tree but as weather prophets and seasonal dominion. In the Balkans, Yugoslavia, and Greece the birds in antiquity were major calendric spirits, just as the same cultures made waterpots and vases in the shape of bear paws in the sixth millennium B.C. In both Asia and America, long before the megalithic seasonal markers like Stonehenge and the numerical charts of the year, bear hibernation and waterfowl migration were the principal markers in the division of the year, "the basis," observes Marija Gimbutas, "for models of a seasonal symbiotic parallel between the social life of men and animals."

Many traditional Siberians believe that birds are the souls of the human dead. In Evink thought, the *beyen* or body-soul of a dead clan member is accompanied by the shaman to the land of the dead at the root of the clan tree or the mouth of the clan river. Meanwhile, the *omi* or shadow-soul flies off upstream to the headwaters, where the immortals live. When Mangi, the bear, chief ancestral spirit controlling entry into the lower world, discovers that the dead person lacks his reflection or shadow-soul, he sends the goldeneye duck to fetch it. The duck fails and the bear itself goes to retrieve the *omi*. The *omi* returns as a bird, flying back to the middle world of the living. The bird-souls are seen there by people, hopping around in the trees before reentering a woman or animal's womb to be reborn. While in the womb the feathers drop from its wings and they become arms.

The Pole Star is, of course, the center of a massive wheel, around which turn the constellations. The power of the two bear constellations is at the hub or source of universal energy, a great nightly chase in the sky. Inverting the probable historical sequence, numerous Siberian tribes explain the bear's presence on earth as a descent from the sky. The Finns, Ostyaks, and Voguls tell a story of the earthly bear's origin on a cloud near the Great Bear constellation, from which it came to earth to establish the Barenfest ceremony, as indicated in the Bear Mother story, then returned to the sky. Like all other bears since then, once killed, their spirit was sent home by the ceremony that the first bear taught to humans.

In any case, people probably learned very early in our species' existence that weather has its origin in the sky. As the great seasonal prophet, the bear would certainly rate a place in that part of the sky which seemed to govern the rest. It is the Prime Mover. In *The Migrations of Symbols*, Donald A. Mackenzie argues that the pagan ceremonies of circular, ecstatic dances were "originally . . . performed by magic-workers to stimulate the Great Bear Constellation." Otherwise, he says, the ancients believed that the Great Bear might jam, or else spin in the wrong direction. According to C. G. Jung, Mithras, the chief divinity of ancient Rome and influence on Christian rituals, is himself the *Sol invictus*, holding in his right hand the constellation of the Great Bear, "which moves and turns the heavens round." By Mithras's time, two thousand years ago, the human heroes had taken over from the Bear Mother.

When the first European whites came to North America they found to their astonishment that the Algonquin tribes already identified the Great Bear using the same stars as they themselves did. In time, they also discovered that this was true of peoples from Nova Scotia westward to Point Barrow and down the Pacific coast, even among the pueblos. The Iriquois and Micmacs regarded the Great Bear as composed of four stars, pursued by seven hunters. A nearby group of stars constituted its den. The hunters were robin, chickadee, moosebird (gray jay), pigeon (passenger pigeon?), blue jay, owl (horned owl?), and saw-whet (a small owl). The pursuit begins in late spring. By fall the four at the end of the line have lost the trail. Robin, chickadee, and moosebird overtake the bear in mid-autumn and kill it, the fall foliage stained red by the bear's blood. Perhaps the four that drop out are the same species that migrate away in winter, while the successful hunters of the clan are those that either stay for the winter or at least the fall.

The great cosmic hunt dominates the northern sky, sweeping counter-clockwise from horizon to zenith to horizon. Ursa Major, Ursa Minor, and Bootes, making up the prey and pursuer, pivot around the Pole Star, a great arm of lights, dominating the northern sky. In the Paleolithic era it was not Polaris but another "north star" that was hub of the revolving universe. The pursuit is like a great gear on an invisible axis, seeming to drive the whole stellar panorama with its energy, bringing the sun, whose brightness is therefore also the bear's doing. It is the ultimate presentation of the food chain, and hunters have always known that the chase liberates the energy that turns the world.

The particular stars and players vary somewhat around the world. In the archaic myth, the big bear and little bear together compose the cosmic elk, Kheglen, pursued by Mangi, the bear spirit, the constellation Bootes. In Europe the bear is not the hunter but the hunted. Along with the little bear, she is chased by the human hunter, Bootes. Among the Algonquin Indians of America, the stars near the pole are the bear's den, the four of the "dipper" the bear, and the seven hunters trailing out behind include part of Ursa Major and four other stars.

The passage of energy, dramatized in the food quest, is the centerpiece of the cosmos. The hunt is not a frenzied pursuit but a stately procession of final things, energy gained and spent, transferred, assimilated, and dissipated, only to be renewed again by the sun. The bear dominated the northern sky as predator in one view, prey in another, reminding us of its high place in the food chains of earth, hunted by humans and yet an avatar of the forces that rule all life and turn the great wheel of the cosmos.

In the Netherworld

As we have seen, the crucial episodes of the Bear Mother story take place in the cave of the bears. It is clear that people have long known that something "goes on" among bears belowground that is different from the other burrowers and tunnelers in the earth. The hibernation of bears must have been one of the great discoveries in the evolution of human thought.

The old cave bears, extinct now but known to Neanderthal people, used natural caves whose passages extended to mysterious depths. Polar bear females prefer to dig through the snow into the earth. Black bears hibernate in hollow trees, but sometimes go underground. The brown bear, which is truly *the* bear of human thought and myth in the north, digs a den among the roots of a tree, spending as much as half its life underground.

In its physiological limbo, it is as though the bear mimics death, going to it readily in the fall as though demonstrating that it is only a journey toward the mother, where life ends and begins. Arising from "death" in the spring, it foreshadows the concept of burial and anticipates the funereal rites by which the human mythic consciousness acknowledges and celebrates the survival of the spirit.

Like the branches that frame the clan tree of life, a profusion of stories sprang from this idea of the bear as the consummate master of regeneration beyond death, or palengenesis, the great round. The bear became not only the model of the nether journey but the keeper of that realm. One story is based on the metaphor of the den and the interior of the body itself. A story is told by the Siberian Evinks:

> A girl, Kheladan, was walking and came to a bear. The bear said, "Kill me and cut me up. Place my heart to sleep beside you, put my kidneys behind the hearth, my duodenum and rectum opposite you; spread my fur in a dry ditch, hang my small intestine on a dry, bent-over tree, and put my head near the hearth." Kheladan did as the bear ordered. In the morning she awoke and looked. Behind the hearth were two children (the kidneys) playing; an old man (the head) slept near them, and opposite were an old man and an old woman (the intestines). She glanced outside; there were some reindeer (the fur) walking about, and the little valley was full of reindeer.

The recovery of good health in the depths of the body, also a form of renewal, belongs to the transformative aspect of the verb *bear*. In healing ceremonies of the Selkup Siberians the shaman, dressed in bear fur, seeks access to spirits of the lower or inner world. Our dreams are also inner phenomena, correspondent to the events of an underworld. Bear dreams, say some psychiatrists, come from mnemonic traces of our birth, and the ancestral home mountain of the Bear Mother story is the *mons veneris*. Dreams of climbing a tree to escape a bear are interpreted as a return to the tree of life or the womb.

Each night in sleep we hibernate a little, a tiny death: our breathing and temperature are modified; ingestion and elimination are forgotten. The bear is at the mammalian center of this excursion toward death's second self, says Carl Jung. Of animals in dreams James Hillman observes, "To look at them from an underworld perspective means to regard them as carriers of the soul, perhaps totem carriers of our own free-soul or death-soul, there to help us see in the dark. To find out who they are and what they are doing there in the dream, we must first of all watch the image and pay less attention to our own reactions to it. . . . No animal ever means one thing only, and no animal simply means death."

"Underground" comes into our own time in a related context—as counterculture or minority. Gary Snyder, in a poem entitled "The Way West, Underground," has reminded us of a deeper version of American history. White Americans trace their roots to the East and the coming of Europeans, but there is a prior history. America was first populated by people from Asia going east, not west, who brought with them the myth and ceremonies of the bear. They, in turn, were indebted to even more distant ancestors, the occupants of the great cave sanctuaries of Europe, in whose Paleolithic painting the divine bear is represented at crucial points in the passages.

Thus the oldest American traditions did not arrive by westward Atlantic crossings but through Siberia and Alaska, the same route by which bears themselves arrived in a series of biogeographic surges. Of this the author of the Bear Claw Press calendar for 1977 says, "The earliest cultures followed the paths of animals: hunters from the Siberian taiga tracked bear and caribou into North America. . . . In retracing the bear's ramblings from the West Coast back around Asia to Europe, Snyder's poem describes a cultural continuity as well, returning us to our own deepest origins. . . . It utterly reverses the European view, not only by reconnecting human destiny with its habitat and fellow-creatures, but by recovering the natural history of Northern peoples and animals, moving together across the land masses."

On the Earth

If bears in the sky or in the underworld seem strange to modern thinking, it is because of the loss of animals in our more mechanical and humanized cosmology. But the bears in the middle ground of "nature"—real, physical bears—continue to be familiar. Informative as it is, however, about the demography, biochemistry, and anatomy of bears, the scientific view of them in the middle world where we live our daily lives is a narrowed version of an older, ethnic perspective.

The Inuit Eskimos, for example, think of the bear as both kin and tutelary guide in the food quest. They say their hunting strategy imitates the stalking of seal and walrus by the polar bear, who moves against the wind, crouches below the horizon, advances only when the prey looks away, pushes ahead a concealing chunk of ice, throws rock or ice, even covers its conspicuous black nose with a white paw. Almost everywhere bears are, people are "guided" by bears to medicinal herbs, the location of honey, and the wild fruit and meat of the season.

In the hundreds of millennia of close attention to bears, they were seen in dramas of life parallel to our own. In this growing awareness of reality as recollection and intent, the marvelous array of animal life ceased to be simply Others as each species took its place as a skilled and talented practitioner. Each seemed to "know one big thing" in the metasocial fabric of the world, even as they came to be regarded as the physical avatar of spiritual presences—deities who were each limited in their particular powers. Few among the animals reflected the human situation, and fewer still seemed wise beyond its singular genius. The bear was the great exception.

Beginning with its physical characteristics—the humanlike foot and dexterity of forepaws, tendency to stand erect, binocular vision in spite of a big nose, humanlike anatomy when skinned, and so on—its major qualification for viziership is a consequence of its omnivorousness, hence its versatility. Omnivorousness shapes the personality of bears, as it does humans.

It is the bear's broad, searching, persistent openness of attention that is familiar to us. Every bear generates recognition of a fellow being whose questing, provocative, garrulous, taciturn, lazy ways, even whose obligations and commitments to cubs, to hunt, to hole up, and to dominate a space remind us of ourselves. It is because of this that the bear is both observed and ritually eaten by humans for reasons beyond food and practical life. The association of the bear with the salmon river and the tree was perhaps the earliest step toward this identity. And it was the tree and the river that connected the three levels of the universe.

Of the watery connections, a Siberian story explains the relationship of the bear to the creation of tools. The myth, as recounted by G. M. Vasilevich, is that "The bear started to cross a river, getting deeper into the water—up to the heel, the ankle, the knee, the thigh, the hips, the belly, the navel, the armpits, the shoulders, the throat, the chin, the mouth, the nose, the eyes, the crown of the head, until he disappeared altogether. Then he said, 'My heels shall be whetstones, my knees grinders, my shoulder blades stones for trying out colors, my blood the color red, my excrement black.' Since that time one could find in the taiga colors, grinders, whetstones, and other things."

Sick, injured, hot, pestered by horseflies and midges, bears go to water. To it polar bears flee for safety and to it the she-bear leads her young for their first seal hunts. Swamps in the American South are common refuges, daybeds, and denning places, for the trees there grow especially big with hollow trunks. Like his little cousin, the raccoon (or *wassbar*) the bear prowls wetlands and seashores, seeking windfall meals.

In the past the rivers of the northern continents teemed with anadromous fish, clear to the centers of the continents, especially the salmon, whose annual runs were as tightly time-factored as the migration of birds or the mass emergence of certain caterpillars. Like the waterfowl that follow the stream in flight, the salmon ascend rivers from their mouths at the mother sea to the tips of their tributary branches, as though climbing the clan tree, bent on a transcendental voyage in which the familiar middle world is but a segment, where it encounters the guts of bears, eagles, and people.

The Greek bear goddess, Callisto, was said by Hesiod to have been a river nymph. Her other name was Themisto, daughter of the river god, Inachus. Her tomb was near a spring at Krounoi in central Arcadia, connecting her not only with a watery environment but to an underworld entrance. The bearish places of preclassical Greece were all near springs, as in the North, where the bear cult pottery of the ancient Danubians was decorated with bear paws and symbols of flowing water.

As for trees, bears at the southern part of their range are born in hollow trees and, in the case of the spectacled bear of South America, make sleeping nests in trees. Tree dens are for sleeping, hibernating, and escaping, as little bears are taught to scamper up trees for safety. All species, except polar bears, climb.

Willow and alder buds, like fruits and acorns, are sought as food. The underbark of fir and spruce is widely eaten. In most places the forest is the bear's habitat. Living out of it or deprived of it, bears travel by barrancas and gullies or between moraines, become more strictly nocturnal, get gruff and aggressive toward people rather than stealthily evasive, eat different food, and grow to small adult size.

Traditional tales link the human use of wood as fuel to the celestial bear, whose pursuit of the cosmic elk keeps the wheel of the universe turning and brings the return of the sun's heavenly fire. A Prometheus-like story of the Ainu tells how the bear obtained fire from volcanoes. The underworld associations make bears familiar with the fire inside the earth, just as their solar connections are implicit in the sun's control of the seasons. Among several Siberian tribes, the bear ceremony includes a procession in which the bones of a bear are carried around a wood fire in a solar pattern.

Perhaps the most incisive connection between bears and trees is the ursine habit of "signing." Pandas spray their urine directly on trees. Balkan brown bears roll in their urine and rub their backs on trees. Rubbing-trees are also marked with scent glands in the face and back. Bears reach up and scratch hemlock, spruce, fir, and pine trees. Individual trees are scratched again and again over a period of years. Whatever the meaning of these marks to bears, humans have seen them as communicative, as art, as signs to themselves.

In bears, deciduous trees have a metaphysical analogy, companions in the art of seasonal renewal who seem to die but green up with life in the spring, when both are discovered not to have died. Like the tree, with its roots, trunk, and limbs in different layers of the cosmos, the bear is seen in the stars in the night sky, underground as the sleeper, and on earth as mentor and food gift. Yet, in ethnic tradition, there is only one bear. William Gronbech, a Danish historian of religion, says, "When the animal steps out of our view we fancy that it trails a line of existence somewhere hidden among the thousand things of the earth until it reappears across our path. . . . The universe is crossed by millions and millions of threads, each one spun by an isolated individual. According to primitive experience . . . all bears are the same soul, and every new appearance of a bear—whether it be no other than that we saw yesterday or the most distant of all among the kin, as we reckon—is a new creation from the soul. A bear is a new birth every time it appears anew, for the deep connection in the existence of the soul is a steady power of regeneration."

So it is that the Mistassini Cree Indians say, "If we do not show respect for the bear when we kill him, he will not return." The crucial element in the soul's return as a body is human participation in its departure. Such were the instructions given to its children by the holy bear, husband of the Bear Mother.

Other animals have tutored humankind in this matter, all having in common the capacity to survive disintegration: the butterfly, who is transformed from the caterpillar; the snake, whose annual reappearance is associated with skin-shedding; the frog, altering from tadpole in the water to adult on land; and the beetle, turning a ball of dung into new life. Translated into the human sphere such passages are symbolically framed, artistically vivified, ritually performed in the transitions of birth-death, sickness-health, childhood-puberty, singleness-marriage, and the movements through social ranks and identities.

The bear subsumes them all, the model of watching and thinking who tells us that those phenomena are the keys on which to construct a life.

Searching Out Kindred Spirits

If civil liberation can be said to have leaped forward in the 1960s, "animal liberation" began its modern resurgence in the 1970s. The impulse seems alike in both cases; the differences are hidden and deep. Both movements share a generous urge to support and honor life, but the moral rejection of racism within our own species cannot be extended to a rejection of death-dealing between species. This would thwart an essential factor of organic existence. Diversity and kinship of life not only include the fact of death but require it.

Causing death—requiring it in order to live—pricks our conscience. In the city, it is possible to hire professionals who hide the death in the slaughterhouse. But in the world at large, the hunt must be faced. The mood of recent history is to celebrate "hunting"—seen in the sense of the search for truth—but to disdain the literal hunt. However, can hunting make sense, but not the hunt? Is the primal hunt, the grand paradigm of all prehistory, irrelevant except as metaphor?

This difficult relationship between physical reality and metaphorical reality is not new. The earliest human art and rituals tell of the quest that grew up around the killing of animals—a kind of double seeking, both physical and spiritual. Men sought an accommodation of meaning as they hunted their food—and sometimes as they themselves died as food for other species.

What can we learn from this perennial search? That hunting is a holy occupation, framed in rules and courtesy, informed by the beauty of the physical being and the numinous presence of the spiritual life of animals. Among traditional hunters, intelligent quarry are the most prized and respected. Animals are not classified into levels of significance that excuse the killing of "lower" forms and preserve the "higher." Instead, it is just those great sensitive beasts—aurochs, horses, elephants, and bears—that are most appropriately killed and most deeply revered. Where we see the consumption of animals as participation in the transfer of energy, as do traditional hunters, they also see it as the movement of an endless spiritual

47

flow. To live and to die is to be surrounded by beings whose coming and going are intrinsic to life itself.

The killing and eating of animals by hunting-gathering peoples is not seen as a victory over a reluctant nature, nor as an assertion of will or virility, but instead as part of the larger gift of life, a receiving from the hand of a conscious power according to the state of grace of the recipients. The crucial moment in the hunt is not the "taking" of a life but the moment of respect and affirmation for a giving world. Nor is the hunter alone considered to be responsible for the fatal stroke. The accountability is group-wide, and the ceremonies that ensue—including rites of purification, celebrations, homage, and veneration—are participated in by all.

In the hunt, we focus our attention on the mystery of life-giving substance—a substance available only in death (whether that of plants or animals). All transformation grows from this one; it powers the changes that mark the stages of life and even the birth of the cosmos.

No one should condone the mutilation or unnecessary punishment of animals. All captive and domestic forms are entangled in a web of social and ecological circumstances that require clear, cautious moral judgment. Even so, "rights" and "ethics" are pale, rational-legalistic concepts. In the case of the hunt, such socially defined concerns are wholly inadequate. In killing, some people try to work down the scale of life as they imagine it— from big, smart animals to the mindless fish—hoping at last to eradicate the butchering of bodies like their own. This frantic perpetuation of life is deeply antiorganic, a denial of our own bodies.

It is, in truth, a denial of every aspect of our nature. We are each porous, actively exchanging with the world at large. What we take in—and not only as physical food—becomes some aspect of the self. Perceptually, animals constitute elements of the potential self as it grows. In their taxonomy, animals provide us with the concrete reality of categories of existence. In our dreams, they symbolize the resources of the self. In their behavior, they model for us—as reflected in our childhood mimicry in dance and play—our vast range of feelings.

The great miscalculation is to say, "Welcome all births, save all lives." Compassionate on the surface, its effect is to destroy life, in the long perspective causing more havoc than does ordinary malevolence. The alternative is not an ascetic otherworldliness that will "deny birth and welcome death." Instead, it is as if one were to say, "Celebrate new life as provisional, affirm that death will balance the scale, and accept that human beings are part of the whole."

Regardless of food habits, everything from protozoa to tigers incorporates other life to live—other life that must be searched out and must die. All are hunted in turn. The great predatory carnivores demonstrate it most plainly, but even they, in the end, are pursued by microbes, fungi, and plant roots.

Humans are uniquely endowed to know this. Every mature culture celebrates it in music and ceremony.

Like all human activity, hunting can run amok—in both killing and estrangement. We all know of the modern abuse of hunting protocol, the secular trivialization of killing. But what is the answer? The humane preservation of wild animals cannot be set apart from the hunt; it requires much mature thought. We are easily seduced by our own empathy because of our fear and outrage at the indifferent destructiveness around us. However, kindness toward animals demands a true sense of kinship. To be kindred does not mean we should treat animals as our babies. It means instead a sense of many connections and transformations—us into them, them into us, and them into each other from the beginning of time. To be kindred means to share consciously in the stream of life.

On Animal Friends

During nearly all the history of our species man has lived in association with large, often terrifying, but always exciting animals.
Models of the survivors, toy elephants, giraffes and pandas, are an
integral part of contemporary childhood. If all these animals
became extinct, as is quite possible, are we sure that some irrep-
arable harm to our psychological development would not be done?

G. E. Hutchinson, *The Enchanted Voyage*

Behind all discussion of the relation of humans to other animals is the
final and irresolvable enigma of our identity—personal, social, and as a
species. This essay begins with the "savage" mode of self-identification
by reference to others and then arcs forward through the end of wildness
to modern narcissism as the failure of such a reference.

Among the charities that impinge on our daily lives are those for
saving, rescuing, rehabilitating, and protecting animals. We are
exhorted to treat them as fellow beings, even as friends, to extend our
horizons to include them within our circle of familial and social responsibility, and to be gratified in this "humane" caring for all things great and
small. Is this partnership what is meant by the "innate tendency to affil-
iate" that defines biophilia? "Innate" implies evolutionary roots and
specieswide characteristics. It evokes the past in the strict sense of our
ecology and the Pleistocene perception of animals by our ancestors,
who lived in small, foraging groups.

Classical scholars of the past often characterized the "barbarian" mind
as capable only of fuzzy distinctions between the self and its environment—the muddled thinking of a tribal horde "at one" with nature. It
appears, however, that primal peoples (variously termed primitive,
savage, tribal, subsistence, hunting-gathering, Paleolithic, indigenous,
and ethnic) elaborate the distinctions between culture and nature—
beginning with that which is human and that which is not—and then

pursue ever-finer refinements. This line between culture and nature, defined by Claude Lévi-Strauss as totemic, creates "a homology between two systems of differences, one of which occurs in nature and the other in culture."[1] He refers to animal species on the one hand and a taxonomy of human groups on the other, and it is evident that the "system of differences" includes ecological and ethological characteristics among the nonhumans and a parallel track of social behavior among the humans.

Contact across this polarized field is fraught with power and danger. Among primal peoples, contact with the living nonhuman world is hedged with circumspection, caution, and sometimes ceremonial formality. Invisible realities play through this concern—animal spirits, sacred beings, and ancestral presences. Since they have no domestic animals, the physical presence of living creatures in close contact is unusual, even though the use of animals as food and skins is extensive.[2] One does not abuse their remains or consort familiarly with such very different "peoples" with impunity, for they are sentient, alive or dead, and can influence human well-being. It is inappropriate to behave in a companionable way with a deer or a wolf, even if the deer is prominent in a material or spiritual sense and the wolf is a fellow-hunter, both major figures in the mythology. As a consequence of this division, reaching across the boundary between culture and nature, the exercise of utility obligates humans to canonical acts of depressurization resembling diplomatic obligations. The hunt and other encounters with animals in daily life are framed in aversion, circumspection, convention, protocol, thanksgiving, and acts of contrition or apology but not as ordinary gregarious conviviality, thoughtless exploitation, or the dominance relationships of slavery. The killing, skinning, cooking, and other uses of an animal are circumscribed by customary, rhetorical practices at once efficacious and mitigating. Reticence marks the contract between human and the Other. Not only the deer but all species are seen in this way, as members of a polythetic cosmos, a community with its own implicit, intergroup formalities, as if the universe were a vast social drama.

This does not mean that the Lévi-Strauss line between the humans and the Others is simply an edge, the crossing of which requires traditional courtesy. It is also the zone of translation between the distinct domains of nature and culture. The ecological side is perceived as a coded reference for the rationalization of human society. It is as though all human societies were endlessly clarifying their internal distinctions. Endowed with the furor and tumult of primate group life, humans relentlessly repair and rephrase the social contract, often with reference to some external model. The naming of clans after animal species is an example. A consequence of this reference is, Lévi-Strauss notes, that culture is not regarded as an improvisation but is perceived from within as "natural." At the same time this allusion bridges the artificial distinction between nature and culture, mending the conceptual damage done by such dualism. If a binary approach to the world makes metaphor possible, it is healed in due course by myths of a shared ancestry of humans and animals characteristic of totemic societies and the perspicacious idea that nature is a language and guide to human life.

The wild animals composing nature are an array of species whose differences and interactions are observed and translated by keen, lifelong attention to the nuances of natural history. Out of this observation emerges the notion of a comity to which human society is analogous. Lévi-Strauss describes wild species as the concrete model of categorical forms, from which societies give names to their clans or other subgroups, each having its eponymous or totemic animal. In short, people justify group relationships by such a poetic reference, giving them expression in narration, art, fetes, culinary life, and intergroup protocol—all out of a kind of logic, not of emulation but of parallels.

If this nature-culture correlation is as widespread among tribal peoples as Lévi-Strauss implies, it is probably extremely old. Some prehistoric art can be fitted to its general framework, suggesting that the system of parallel worlds might very well be basic to the evolution of human consciousness. A possible stage in the evolution of ecological patterns as poetic models of a human social matrix may be what Paleolithic cave art was about—that is, as the middle period between the earliest symbolic reference to animals and the full flowering of cynegetic thought at the end of the Pleistocene. Bertram Lewin argues that cave art may represent the deliberate effort to collectively internalize key images in a group of observers, perhaps initiates. The darkness of the cave is experienced as a shared reference to the darkness of the cerebrum, where in imagination we glimpse figures, as it were, in a brief light—a mode of bringing a fauna into the head.[3]

Regional differences in such practices may have tended to isolate human groups from one another, leading to cultural differences such as the emergence of linguistic dialects and the islanding and evolution of races, giving cohesion and identity that bind members to the group or set groups apart. If this nature-culture line characterizes the evolutionary origin of human cognition, a root of biophilia, it is in us still, the tug of attention to animals as the curved mirror of ourselves—not as stuff or friends but as resplendent, diverse beings, signs that integrity and beauty are inherent in the givenness of the world.

Minding of other animals is far older even than the human arts of the Paleolithic, however, and is widely shared. We are not its sole heirs. Harry Jerison has described the Cenozoic evolution of mammalian brains, notably the savanna predator-prey system in which the attention structure was focused on other species in the mutual honing of pursuit and escape strategies and the reading of signs, resulting in progressive, reciprocal enlargement of the brains of hunters and hunted. The advent of protohumans into this savanna game as part-time carnivores and occasional prey corresponded, he believes, with the emergence of speech and the cognitive need to determine what constitutes "objects" or "categories" in the perceptual field to which words could be attached and which thereby become real. In short, as we became open-country hunters and hunted, we entered an ongoing system of brain-making, using our advanced primate vocal and visual apparatus instead of an olfactory system. This process centered visual imagery on the simultaneous and tireless scrutiny of other animals and the emergence of self-consciousness.[4]

The metaphysical role of animals in primal societies that shaped and defined our species is not just a vague "reverence" for animals but a body of procedure and narrative that acknowledges kinship and the necessity of killing, a sinew of sentience and spiritual power linking death and love. The Pleistocene offers us no compassion for animals in the warm idiom of the teddy bear. It constitutes instead the genesis of self-consciousness as assimilation, an endless scrutiny that is both instrument and synonym of becoming—the ecopsychology of predator and prey. Eating, intaking, is the culmination of the holy hunt, a sacred meal in which not only energy but qualities are internalized. This was the new integration of an anthropoid mind with savanna thinking. And it continues to remind us of the reality and significance of wildness in our individual becoming and our extraordinary origin in the game.

As its strategies became conscious our savanna-based, hominid, omnivorous ancestor assembled the self cognitively by reference to an extensive fauna swallowed bit by bit. As E. H. Lenneberg recognized thirty years ago, semantic structure is native, while vocabulary is acquired. But both depend on a given world. Category making relies on external prototypes.[5] Inherent grammatical structure articulates with a perceived array of species. Human development—as a result of two million years of participation in the savanna game—endows each of us with an epigenetic ontogeny, personal development combining inherent predisposition and the right experience, in which speech and cognition are keyed to the natural world.

This is why awareness of animal taxonomy is so prominent in childhood. Species recognition has been described by Eleanor Rosch and others as a two-step process in which the whole animal is first identified in a major category, such as "bird," and then defined to kind, such as "sparrow," by specific cues.[6] Such cues are compelling because of the early attention to body parts or, one might say, the "butchering perception." Everybody is familiar with the infant's interest in external anatomy, its shifting attention between the other and the self in an association of tactile and visual play—the touch and say "eye," touch and say "nose," touch and say "ear" with a caregiver. It is antecedent to pointing. The reciprocity between infant and mother in these rituals of naming and touching toes, nose, fingers, and belly is perhaps the procedural model for a lifelong process of naming based on distinct, detachable characteristics. For these reasons the principal nouns of interest to small children are body parts and animals.

Life is centered on the enigmas of *I, we, you,* and *they.* The pronouns are conceptually slippery. As a group or clan member, the self could be identified as part of a social species in the way Lévi-Strauss suggests. But we know the self from an inner life as well, which is more obscure and more personal. One's own "body percept" is built up incrementally, first outer and then inner. Our external anatomy has corresponding parts among animals, while our inner anatomy also matches the body parts of nonhumans. This physical correspondence is a kind of map of feelings. The visceral terrain is composed of felt events. In their organs and behavior— and thereby in their names and images—animals provide the concrete reference for sorrow, pain, moods, tempers, dispositions, and all those vivid, chromatic tides that wash through our being, which we know as our own life. In this way animals fill the void to which our self-consciousness would be subject without prismatic representations of qualities or bits of experience. This discovery of the external existence of animals who correspond to our inner reality began not as an invention but in the natural history and evolution of mind. Both character-filled story and anatomical taxonomy depend on the close association of speech with the early naming of body parts and actions, followed in childhood play by the mimicry of animals. From external morphology to the identity of internal organs (mythically animated in some people as having lived independently like other animals) to the less tangible parts of a self, there is the progressive construction of the world as pronouns.

The various steps by which we discover the inner animals have both phylogenetic and ontogenetic aspects. Verbalizing the names of animals may have come early in the evolution of speech and is a stage event in infancy and childhood in which the creatures are swallowed by attention—as though humans had not been convinced in their nascent carnivory that the animals eaten were not still somehow alive inside. The verbing of these nouns—the vast body of verbs, gerunds, and infinitives: to cow, fish, duck, quail, clam up, skunk, weasel, outfox, hound, dog, goose, horse around, lark about, hawk, worm your way, bug, ram, pig out, hog, grouse, fawn, buffalo—constitutes the lexicon of an inner structuring dependent on a system of referent species whose actions are a speaking of individual experience as a fauna. Having recognizable external parts resembling my own, animals are an invitation to think about my own visceral landscape, of our corresponding organs and behavior, so that the fish seems to embody all random seeking and searching in our lives, as though created to present us with that particular representation of ourselves.

In the human developmental calendar the animal is first a noun and then is verbed not only in speech but in play. Stylized and conventional movements (prefiguring ritual in adult life) are essential to play—in which we find the "predication of the animal on the inchoate pronoun" in games such as Sharks, Wolf and Sheep, or Fox and Goose, as the participants define themselves briefly and successively as different animals.[7] The enactment of the verbed animal captures the evanescent feelings that go with "wolfing" one's food or "chickening out" of a dangerous situation.

Such games are like enacted stories. And among these stories is a genre with happy endings in which the normal concerns of childhood, personified and distanced from the self, are defeated in the end, often with the help of animal figures from deep in our organic substrate. The personified feelings or states that the protagonists in fairy tales embody in listeners include different animals representing slightly disguised (that is, perceptible to the unconscious) aspects of the self as a kind of promise. The hearers of stories that begin "once upon a time" and have a happy end ruminate on the tales as they foreshadow intrinsic solutions to childhood worries and register a deep correspondence based on the double life of the protagonists—for example, the dove as one's spiritual side, the wolf as aggression, or the frog as the inevitability of metamorphosis and growth.[8]

In yet another mode, adults can visualize and summon the animal avatars of their own diverse physiological and psychological interiors as aspects of experience and personality. This mode has emerged recently as a distinct therapy. First with a guide and then independently, one meditates on an encounter and imaginary conversation with a series of inner animals who emerge from their places in the vital (chakra) centers or from the bodily organs to speak of their needs and fears.[9] The human capacity to generate these images, under circumstances less focused than hypnosis, testifies to a facility too easily characterized as archetypal and too casually dismissed as imagination. The train of fictive interaction between these animals and their human host is astonishing evidence of the fundamentally zoomorphic perception of one's inner life—a domain immensely troubling and destructive since the reign of "axial man" sought to banish its numinous presence and mechanistic biology reduced our concept of the organs to that of a vegetable-like stasis. The interior fauna reveal a realm of dynamic animal mediators who both embody and represent that which is otherwise obscure to the conscious self. They confirm the self as alive within—a community of being congruent with the outer, living world and dependent on it in the cognitive development of the individual.

These three modes—story, animal games, and visual imagery—are the heritage of two million years or more of the assimilation of animals in diet and mind. It should be added that such self-realization is still an enigma. The awakening of the mediators requires devoted attention, for animals are more complicated and interesting than we think. There is a wildness about our own feelings that they embody—a wildness resistant to final capture, a strangeness which is itself an aspect of being, a hidden side to our Otherness that is like the creatures of an ecosystem who remain underground.

The foregoing may be represented by the schema of the accompanying Figure 6.1. Beginning with Lévi-Strauss's simple diagram, I, at the top of the figure, I have elaborated this binary culture-nature system in II and further in III. Part 1 at the bottom shows Lévi-Strauss's totemic patronymics derived from species A, B, C, and so on. Parts 2 and 3 at the bottom show that emotions and other features of the external world that need "organizing" may also be perceived as analogies to the animal taxonomic system.

If the development of the person's sense of his or her own structure depends on the beauty, strangeness, and diversity of a wild fauna, assimilated ceremonially as food and perceptually as the plural assembly of the self, what can the collapse of the Lévi-Strauss line mean as animals are domesticated? Very few species cross the borderline to live in the human domain. They are like refugees from a ruined nation or guerrillas in support of a failing ministry. Once across, captive and bred, domestic animals become numerous, docile, and flaccid, their brains diminished, their anatomy and physiology subject to dysfunction, and their ethology abbreviated. At the same time the remaining wild fauna recedes from human sight and table. The anthropological literature bulges with human attempts to make do cognitively with cows, pigs, horses, dogs, and chickens, their impoverished zoology and hybrid extravagance.[10]

Domestic animals (and plants) are the products of the first genetic engineering. They have served for millennia as material substance and workers, though celebrated in principle as sacred gifts and epiphanies. In recent times their role has evolved from barnyard utility to companionship. As pets they contributed to human well-being long before anyone asked why. Now there is statistical evidence from therapeutic programs proving domestic animals to be anodynes for human suffering ranging from penal incarceration to Down's syndrome. No one who looks at the evidence can doubt that animals in hand improve the quality of modern human life, whether measured in terms of longevity or recovery. Pet keepers can take consolation in knowing that their animals are proven health additives. The efficacy of this dumb-beast panacea has an impact comparable to that of antibiotics. Animal-facilitated therapy cuts a swath through despair, loneliness, genetic impairment, the terminal blues,

I. In *The Savage Mind*:

"Totemic Institutions"

NATURE   species 1 = species 2 = species 3 = . . . species n

CULTURE   group 1 = group 2 = group 3 = . . . group n

II. As seen under the hand lens:

$$A = B = C = D = E = F = G = H = I \quad \text{species, the paradigmatic categories}$$
$$a = b = c = d = e = f = g = h = i \quad \text{human social groups}$$

III. As seen under the microscope:

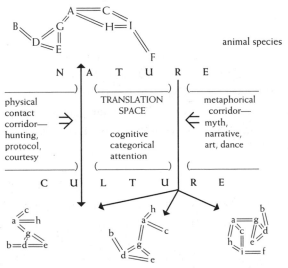

animal species

N A T U R E

| physical contact corridor— hunting, protocol, courtesy | TRANSLATION SPACE

cognitive categorical attention | metaphorical corridor— myth, narrative, art, dance |

C U L T U R E

1. Totemic society (clans, sex, rank, age, kinship)

2. Internal structure (organs, feelings, emotions, other intangible elements of self)

3. Other analogies (space, zodiac, tools, terrain, deities)

FIGURE 6.1. *The Lévi-Strauss or Nature-Culture Line.* Part I reproduces the nature-culture diagram exactly as shown in Lévi-Strauss's *The Savage Mind*. Note that the analogy is between the signs (=) designating relationships. It represents in an elemental way the basic use of an external model for cultural analogy. Part II emphasizes the one-way flow of this connection—entirely as a mental function of humans—and suggests that there are many species. In part III the concept is elaborated. The species relationships are shown in constellated rather than linear form. The boundary between nature and culture is punctuated by two types of connection: the one on the left involves the physical use of animals; that on the right indicates metaphorical use that breaks into three categories. Number 1 at the bottom shows a pattern of clan relationships representing a portion of the animal ecological pattern as given paradigmatic application to human group membership. Number 2 is another pattern, also "taken from nature," indicating internal psychological structures that parallel animal interactions. Number 3 suggests some other analogical uses of the ecological system of concrete images as the means of giving coherence to an otherwise inchoate structure.

pain, the black holes of autism and schizophrenia, the chutes of age, boredom, and immobility. As medicine nips the usual killers in the bud and people live longer with their geriatric syndromes and other civilized maladies, the potential for animal companion tonic increases. Apart from helping the sick and disabled, pet therapy has created jobs for health care specialists, pet mortuary and bereavement experts, grooms, keepers, handlers, new categories of veterinary specialties, breeders, middlemen, the makers of pet food and clothing, designers and builders of special facilities, along with a vast corporate, academic, and medical contingent, and acquires an easy marriage to the animal shelter industry. There is both altruism and financial profit. And the alliance of pet keeping and corporate medicine does not end there. It generates moral judgment, as well, and facilitates animal rightists, animal ethics theorists, antihunters, vegetarians, and the rising tide of fringe animal lovers right down to the zany keepers of legions of cats and dogs in their bedrooms who might otherwise kill themselves or burn down city hall.

From this heterogeneous multitude there emerges a great cloud of righteous socializing with the entire animal kingdom—projecting onto wild nature unlimited fraternization with its baggage of care, compassion, and kindness and its overbearing social injunctions, ideal personal standards, and other traits and rules that shape human society: in short, the invasion of the ecological world in a spirit of human arrogance.

What can be said of the existential status of the institutionalized domestic animals as well as the family dog and cat? Clearly they are happy friends of humankind—just see how they wag their tails and purr. Indeed, they *choose* this life, they prefer it, and they are far better off (as all slave keepers have said). In fact it is part of nature's grand plan. I offer two recent examples of this thinking from opposite ends of the spectrum, the popular and the scientific: a journalist writing of the "covenant of the wild," arguing that domestication is just recent evolution, and the editor of *Science* saying, "Over evolutionary time the friendliest of wolves (and possibly the most intelligent) learned that wagging their tails and delivering slippers was an easier way to earn a living than hunting caribou in the wilds. . . . An original wolf might say to the dog, 'You have lost your freedom. Your obsequiousness is humiliating to the family of Canidae.' The dog could reply, 'I am much less warlike, far more altruistic, and besides, it's a wonderful standard of living.' Whether society prefers to have wolves or dogs remains to be seen."[11]

The editor of *Science* is not just being down-home with his tail-wagging and slippers bromides. Even schoolkids know that wolves did not *decide* to become dogs. The part about warlike and freedom and standard of living exploits our bias in favor of dogs. It is a lie. As for subservience, the worst thing about the editorial is its breezy participation in the larger technophilic fiction that everything is getting better and better through the control of plants and animals—an attempt to co-opt the whole momentum of evolution in the advocacy of genetic engineering.

To understand what the collapse of the Lévi-Strauss line means, we must ask: who are these animals that, domesticated, came to be the model images for all animals? Two biologists, Konrad Lorenz and Helen Spurway, were not so foolish as to confuse evolution and domestication. She, a geneticist, referred to domestic animals as "goofies" because of their addled genetics and the resulting phenotypic mélange.[12] Lorenz made the comparison clear: the domestic had been shorn of what was subtle, complex, and unique in the wild ancestor, including the intelligence and independence meant by "wildness."[13] He loved dogs, but he knew them to be gross outlines of the wolf. With diminished brains and congenital defects, these abducted and enslaved forms are the mindless drabs of the sheep flock, the udder-dragging, hypertrophied cow, the psychopathic racehorse, and the infantilized dog who will age into a blasé touch-me-bear, padding through the hospice wards until he has a breakdown and bites the next hand.

If domestication breaks up wild genotypes, which are continually honed in a kind of DNA harmonic with the environment, it is hardly appropriate to say that the animal is "adapted" to human care. This bottleneck breeding, with its catastrophic breakdown of genetic equilibrium, releasing broods of monsters, is no "evolutionary adaptation." From the standpoint of their relationship to people it is hard to know what to call this protoplasmic farrago of dismantled and reassembled life. All of those living animals drawn onto this side of the nature-culture line enter the human social system—as healers, friends, siblings, offspring, competitors, entertainers, caricatures, sexual mates, bodyguards, protégés, saviors, litigationists, princes, and fairy godmothers, not to mention their participation in the social categories of the homeless, the unemployed, the sick, the deserving, and so on.

In a radio call-in program I referred to animal companions as "slaves." The telephones buzzed with angry callers who insisted that their dog was a comfortable, privileged family member. Their reaction was like that of certain pre–Civil War southern plantation owners who could point with defiant compassion to their grateful, singing cotton pickers. The truth is that pets are subject to their owner's will exactly as slaves. Yet the term *slaves* may not be suitable, since human slaves can be freed by political and social action. The goofies, congenitally damaged, cannot. If freed they die in the street or become feral liabilities. We could simply quit breeding them. Their relationship to us is not symbiotic, either, or mutual or parasitic. None of these biological terms is suitable to describe organic disintegration in a special vassalage among creatures whose heartwarming compliance and truly therapeutic presence mask the sink of their biological deformity and the urgency of our need for other life.

Less than kindly euphemisms for "companion animals" come to mind—crutches in a crippled society, candy bars, substitutes for necessary and nurturant Others of the earth, not simply simulations but over-refined, bereft of truly curative potency, peons in the miasma of domesticated ecosystems. The corporate takeover of the pet is merely a recent step in institutionalizing, rationing, and marketing these ameliorations for something essential and missing in human health.

My concern here is not the destiny of these lumpish, hand-licker-biters among humans who are desperate for the sight of nonhuman creatures because they touch some deep archetypal need. Nor is it the monumental logistics of the forty million dogs and forty million cats in American homes, not to mention the rest of the world, and all the animals in the world's zoos. My focus is the effect of the replacement of domestic for wild animals in our psychological development, especially in the formative processes by which we mature. The colossal upsurge of the pet as an industrialized healer brings the issue of our inner life before us, along with the planet's diminishing wild abundance and diversity. The agro-urban world replaced a way of life centered on the elegant courtesies of totemism and the brain-making hunt, its roots deep in the Pleistocene and deep in the human heart. When animals as domestics came literally into our households across that Lévi-Strauss line, they filled the lowest ranks of our society. There was the end of respect for the Other on its own terms.

The projection of the domestic milieu has infected our perception of all animals. The breakdown of the Lévi-Strauss line results in the impaired ability of the ordinary person to make critical distinctions between denatured goofies and wild animals. The idea of responsibility for the animal kingdom as a whole is clearly neobiblical, especially "care-taking" and all its benevolent expressions. These are three: the Noah Syndrome, which puts us in charge (as God's steward) of *all* the animals; the hagiographic model of Saintly Hermit, before whom the beasts, recognizing human holiness, gladly enter into cringing servitude; and the Peaceable Kingdom, the prototype for our perception and regulation of nature as if it were a nursery school playground.

To recreate the Peaceable Kingdom on earth requires our invasive acts of protecting the weak from the strong, the imposition of interspecies ethics, the infusion of intention such as kindness or mercy, the projection of the domestic world onto nature. Our Noachian authority requires that we choose which individuals shall breed to survive the catastrophes we have brought upon the wild world, followed by their incarceration in the zoo cells of the modern ark. Our saintly status of righteousness and dualistic thought justify these extensions of our power over the wild Others and the expectation that they will submit to our moral eminence. All three concepts take wild animals one step closer to becoming slaves along with their domesticated cousins. Wild animals are not our friends. They need protection from those who, with good intentions, would harm their biology to save them by extending our obligations of care to them, along with its trail of ethics. Paradoxically, the most characteristic feature of modern "animal rights" is its withdrawal. It smells of the armchair philosopher who, as insulated from nature as possible, advocates "letting them be" as a moral stance. It seems to contradict the notion of benign care. But it is merely the final hubris, the sanitized isolation from those whose gaming made us what we are.

Across the fence of the Lévi-Strauss line, tribal peoples see the ecosystem as a code and mnemonic device for the organization of society. With the collapse of that line, the human social system becomes the standard for managing the wild world and reducing the Others to individuals. Lévi-Strauss has described intermediate cultural practices between the totemic and caste cultures, for example, one in which the members of clans—such as raccoon and bear—are expected to behave like raccoons and bears. Although Lévi-Strauss does not define domestication explicitly as a breakdown in the totemic, metaphoric system, it is clear from his study that in "caste" (or domestic) society metonymy permeates and disintegrates an older structure the way mycelia of a fungus leave the outward form of its host empty of its original function. Domestic culture replaces totemic or analogous thought with physical conjunction and leads to literal-mindedness. As Lévi-Strauss shows, clans identify themselves by an eponymous animal. Intermediate societies, becoming domesticated, cease to ponder the animal as heuristic and begin to imitate it. On the face of it, behaving like a bear or raccoon may not be so bad. But when the pig and the dog have become the animals of reference instead of the bear and raccoon, the animal as model of human degradation cannot be far off, and the "beast" in Mary Midgley's sense is born.[14] Thus does civilization slander savagery by transforming its elegant meditations into gross actions. Indeed, the antinomy of attitudes toward dogs throughout the world displays the contrast of the beauty of the wild canid with its domestic shadow. By human standards the dog is incestuous, shameless in its excretory habits, and evil as a latent killer of sheep and among humans as a rabid brute, yet it is admired for its helping habits as herder, hunter, and protector. The dog's modern incarnation as individual personality has not reduced its ambiguity. In any issue of *The New Yorker* magazine you will find dog-man cartoons that reveal the confusion and angst incident to the fuzzy boundaries of identity among an educated elite who expiate their stress as humor.

From this metonymic stew of the animal as friend and object emerges the paradox that primal peoples kept their distance from animals—except for their intaking as food and prototypes—and could therefore love them as sacred beings and respect them as other "peoples," while we, with the animals in our laps and our mechanized slaughterhouses, are less sure who they are and therefore who we are. The surprising consequence is that "nature" is more distanced, not less. Lévi-Strauss diagrams this domesticated situation by enclosing the animal species and the human group in a box to indicate that the relationship is one between terms rather than an analogy between two sets of relationships. Difference, he says, has overwhelmed the connectedness otherwise drawn from a poetic translation of concrete, ecological relationships. Lévi-Strauss's language has beleaguered generations of students—the "paradigmatic" and the "syntagmatic," the "metaphor" and the "metonym." But we can understand these terms as "figurative" and "literal" and recognize the breakdown of the distinction. If we in the pursuit of progress are destroying species, it may not be due only to the effects of industrial, multinational, corporate greed but to our unconscious resentment against the animals themselves, whose analogical and perceptual roles—anticipated by our human nature—are no longer given credence. The animal, refusing eye contact with us in the zoo, seems to convey a final insolence and abandonment in which we, mistaking who has done what, feel ourselves to be forgotten. This situation is schematized in Figure 6.2 to show the incorporation of a few species (species B becomes B', C becomes C', and so on) as they are shunted by way of a syntagmatic domestic ecology into the human social order (a', b', and so on), as objects and surrogates. Some cognitive and metaphorical categorical use of the species system continues, such as athletic teams, but generally the civilized world uses other, less vital, less definitive, and less heuristic systems for defining human groups or other nebulous complexes.

* * *

The puncture of the Lévi-Strauss line by literalness heightens our sense of anguish about nature and its destruction, but the final loss is our own. Identity is the issue—our difference from and likeness to the nonhumans. According to Julia Kristeva, the issue in civilized society is characterized by infatuation with the self. The Greek myth of Narcissus was an attempt "to tackle a problem that the ancient world had not solved—otherness."[15] The "murky, swampy, invisible drama must have summed up the anguish of a drifting mankind, deprived of stable markers."[16] Narcissus, a hunter, having rejected the love of Echo, pauses to drink while hunting and falls in love with his reflection in a pool. Ultimately he "discovers in sorrow the alienation that is the constituent of his own image"; he commits suicide and his substance is transformed into a flower. That is, he regresses into a vegetative state where the questions of "psychic space" do not arise. His anguish reflects an immense loss in which Narcissus "no longer has the thinking *nous* of the ancient world that would have enabled him to approach the other as plurality, as a multiplicity of objects or parts."[17] Kristeva's insights are worth exploring, although she does not seem to realize that the story is not actually about totemic culture but a myth, projected back upon hunters, about despair by urban societies.

What was that multiplicity of which Kristeva speaks? Narcissus's inner vacuum, the empty psychic space, is a result of the disarray and the paradox of the solitude of cosmopolitan life, the doubt, disintegration, and estrangement of the soul in the urban stewpot, the deracination and cacophony of the city, a dilemma remaining with us still: a "falling apart . . . for which the present-day equivalent would be advanced mass

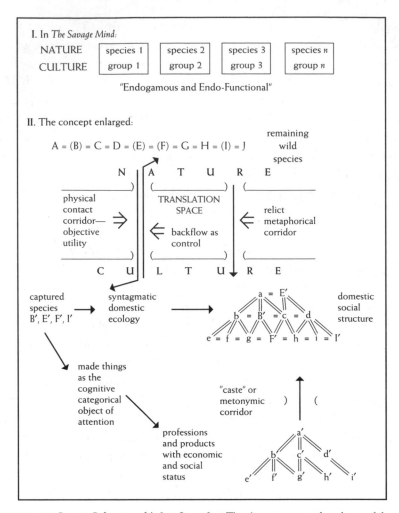

FIGURE 6.2. *Domestic Deformation of the Lévi-Strauss Line.* This diagram repeats the schema of the previous figure with changes brought about by domestication—which Lévi-Strauss labels "caste." Part I is Lévi-Strauss's diagram. He emphasizes the locking of species 1 to human group 1 by boxing them together. No longer does the relationship between species signify as in totemic thought. In part II, the concept is enlarged. Species that have been domesticated are shown in parentheses, although their wild ancestors may remain part of nature. At the left the physical use continues, although it is no longer characterized by formal protocol. The reverse flow along this same corridor, from bottom to top, signifies the projection of human social ideas upon the species system, imposing new rules on the animal ecology. Some residual, metaphorical uses are represented by the arrow at the right. The captured forms have been brought across at the left and domesticated, becoming, at bottom left, part of the system of production—the "things" that have become the concrete models of categorical thought. These are associated with the social class system that emerges from the value and power of the professions, taking a pyramidal form at the bottom and, finally, a domestic social structure. The wild animals are no longer a point of reference, having been replaced by the various products of society. The domestic animals are members of the society, ranked among people rather than studied as model peoples.

mediazation."[18] Seeking origins, she refers unfortunately only to the archaic world of Homeric Greece and the fading of the goddess, and in doing so falls short in chronological explanation.

Kristeva is interested in the recovery of the feminine (and, by implication, the goddess), but neither the goddesses nor gods of archaic history were themselves the primal Others. She notes that the cultures of the ancient world (early civilizations) had not solved the problem. The reason was, in fact, that they had created it. The epiphany of the goddess may have indeed represented a lost sensibility more organically tuned than that of the modern world. But what it replaced was a pantheon of wild Others, the many-specied consort of preagricultural humanity, part of the genesis of imagination itself, a company in whom both the condition and fulfillment of life were incarnate. The nature of the Others was a matter of consuming interest, not a Pleistocene "problem." It was only after the defeat of that numinous, nonhuman presence of animals as a meditation on the nature of the self, in the era of the "ancient" world of cities, goddesses, and gods, that Narcissus rises and falls. All the humanized deities were insufficient substitutes for a zoological theriophany. As the subject and object of its own meaning, the human figure produced disillusion and inner crises, the dead end of remaking the gods in human form.

Subsequently there have been three expressions of the relationships of human and Other within the major Western tradition: Gnostics, who reduced the Others to personifications of evil after the Fall; Christians, preoccupied with the godlike self and the difficulty of integrating it with the singular Other as Kristeva says; and the Neoplatonic Cartesians, for whom the Others were reduced to simple, finite blobs.[19] These solutions to the "problem" of the dynamic connection of the self and the Other have since occupied much of philosophy, narcissism having become a major trait of Western subjectivity.

Plotinus, Kristeva notes, rehabilitated the Narcissus myth for Christianity by calling upon "dignity inherited from ascetic solitude" and identifying the mirrored Other as God. The object was to love one's self, since one is created in the image of God, as the first step toward loving God. Bernard of Clairvaux further argued that Narcissus and his image were the corporeal passion and the spiritual passion, respectively, two aspects of the self, keeping "the flesh from a spirituality that would thus become too ethereal, without forgetting, nevertheless, the presence of a spirit in a flesh."[20] Not that Bernard saw any other saving grace in the organic world or the body: "Be ashamed, my soul, for having exchanged divine resemblance against that of beasts, for wallowing in the mud, having come from heaven."[21] Narcissus, according to this thesis, was merely the failure of true self-love as a fusion with God. If the love of God was inseparable from a love of the human figure, God's image in the self should be One. It had nothing to do with nature—that scorned, plural realm of the beasts. This "privilege of self-love," which began as an Aristotelian idea, was formulated by Plotinus and Christianized by Bernard. The church made a parable of narcissistic despair as a failure of the spiritual imagination to accomplish the necessary unity with God as One. Man, "seized by an unnameable solitude, was called upon to withdraw into himself and discover himself as a psychic being. . . . Henceforth, there is an inside, an internal life, to be contrasted with the outside."[22] "The loving soul must therefore give up its otherness and give in to the sameness of a single light where it loses itself as other . . . as nonbeing." And so, "otherness disappears when we merge with the One."[23] The natural Others—those diverse beings who previously had been thought to be coded clues to the nature of the inner life—disappear or become irrelevant or destructive.

As the church tailored the myth to represent a failed meditation on the unity of human and God, all beings except the human became spiritually irrelevant, leaving the external world as evil (in Persian and Gnostic thought) or available as neutral material for the pragmatic followers of Francis Bacon and René Descartes to do with as they would. The heroic notion of *ego cogito* leads to "the conquest of the outside . . . the outside of nature, to be subjugated by science."[24] Neither the Christians nor modern theorists have recognized that Narcissus begins, not with the loss of the anthropomorphic deities, but with the loss of diverse, therioform Others.

The loss of the wild Others leaves nothing but our own image to explain ourselves by—hence empty psychic space. Portraiture, for instance, arising late in history, is narcissism magnified: the reassertion of the human primate's obsession with the visage. Portraiture presents us with the double bind of a creature that gained in evolution the means of knowing the self as a polythetic being and lost it in repudiation of the biodiversity upon which it depended. The magnification of the face, be it a Rembrandt painting or the latest tabloid photo of a pop celebrity, must, we insist, reveal the person. The art of caricature arises along with portraiture as an aggressive and cynical mockery of this frantic pursuit in which individuals are assaulted by deriding their appearance in terms of bestial qualities that have been made to seem ridiculous and banal after ten thousand years of life with domesticated animals. As William Gregory points out, the closer you look at the face the more animalistic it becomes because of the shared bony structure among the vertebrates.[25]

Mirrors fail us. They reflect that image whose significance is the problem, reminding us of the discomforts of puberty. Adolescent interest in personal appearance is part of a normal ontogeny. In the tribal world the impulse is redirected toward initiation in clan membership—into sacred stories that include composite animals as the instrument of relating disjunct ideas and the ceremonial use of animal masks, giving access to the multiplicity of the self and to cosmological diversity. In a world of mirrors, however, the reflection degenerates into an assessment of our vertebrate skull or the scars of life. Narcissus represents the story of the failed attempt in the quest to know the self by substituting the human image for the mosaic of a wild fauna. It was not, as the Greek myth would have it, the mistake of the hunter and totemism at all. That was a false attribution of civilized problems onto the "savage," a projection by the suffering, civilized mind.[26]

The psychiatric concept of countertransference throws light on Narcissus and the lost role of the Others. An analyst may project elements of his or her own psyche onto the patient and then read them as if they were the patient's. By contrast, when animals are the objects of rumination about the self, they reveal shared elements in the context of Otherness. There is no countertransference—no ricochets like the passive surface of the pool (or the ogled looking-glass eye of a French poodle) from which Narcissus got back, to his grief, only his frustrated simulation.

No wonder the chimpanzee is such a problem. In our Pleistocene psyches we know that its image is not trivial. In spite of our post-Christian, Neoplatonic, and Cartesian self-esteem, we cannot quite dismiss the primates. Their appearance infects and subverts conviction. They seem to combine the mirrored reflection and the intransigent persistence of the Other. A colleague, living with a family of Pennsylvania Hutterites, a branch of fundamentalist Christian Amish, went with them to the zoo, where they refused to look at the monkeys. They knew the trap similitude set for their dogma when they saw it: they felt the numinous power of the monkey mocking their belief that the true god looks like a man. They could repress turtles and birds, but the assault of such likeness was too much, for it carried the full weight of archetypal recognition. This chimpanzee confusion is only the most incisive instance in which we are unable to look away and yet unwilling to accept that in all animals we are seeing aspects of ourselves through a compelling, uncanny lens.

As for the Baconian-Cartesian notion of animal as mindless, to which animal protectionists often refer as the root of the abuse of animals, it was that objectivity after all which led to the laboratory experiments that in turn triggered the humane movement. Perceiving the animal as a machine does not make it hateful—it took the medieval Christians to do that—but merely reduces it to a utility. The medieval abuse of cats, owls, and toads was the result of their being loaded with demonic powers by evangelists who allowed the pagans to keep their spiritual animals in these degraded forms.

* * *

The foregoing argument raises two important questions. It may be asked first whether I have not confused the use of imaginary animals as important mental furniture with real forms. Nothing has been said here about going outdoors, wading through swamps and thickets, or watching grasshoppers. The animals in the psychogenesis of an internal self are those of the mind's eye, not the field. One may wonder whether simulations and virtual realities might not be sufficient. After all, what children actually experience is mostly pictures, toys, films, stories, and other representations. Even the crucial ceremonies of theriology among tribal peoples are centered on storied, drawn, sculpted, and costumed representations. If the occupants of future starships are to spend years away from a wild biota and scatter human colonies on moons without wild species, would not some good holographic animation be more to the point and easier to take along? What has all this development of the inner self to do with actual wild animals?

The answer involves the ecology of maturity. The developmental process is not an end in itself. Cognitive taxonomy and artifacts are indeed the tools in the perceptual work by which the whole person is achieved. But the effect of a healthy identity and maturity is realized in attitudes toward the environment, a sense of gratitude more than mastery, participation in a rich community of organisms, a true biophilia or polytheism. The images—animal guides and mediators—are the representations of an outer world that made our own being possible and toward which our maturity has its end: the preservation of the world. The obligations of having evolved in natural communities constitute a kind of phylogenetic felicity in which we acknowledge that the fish, amphibian, mammal, and primate are still alive within us and therefore have a double existence. They are present as bits of DNA, affirming kinship, and also in the world around us as independent Others. The concept of biodiversity as a social value grows from an inner world and creates respect for a mature ecology, that is, "climax" ecosystems with their diverse inhabitants.

The second question concerns the apparent adoption of a dualistic mode of thought in the nature-culture division. Binary distinctions are necessary for the development of the cognitive skills of categorization. But a holistic or truly ecological perspective should transcend and supplant such schemes of oppositions. Dichotomy, however, characterizes much of Western thought as a full-blown ideology, lending power to modern social and ecological imperialism. The wisdom of the nature-culture line can only be defended as a preliminary step in a process of recognizing, clarifying, and naming plurality and ambiguity—to be succeeded in more mature reflection by a structure of overriding interrelationships. At the end of *The Savage Mind*, Lévi-Strauss observes that the division is merely methodological.[27]

Kinship is the transcendent issue of maturity because of the necessary equilibrium of likeness and difference. Immaturity perceives a world of unresolved ambiguity and contradiction, as though a plurality of powers (and the multiplicity of the self) were defects. What the mature self (and Paleolithic culture) understands is that ambiguity is an intrinsic characteristic, not a deformity resulting from a Fall or a dialectical problem. Such misinterpretation marks an estrangement from Others, no longer balanced by the sense of generic relationship. Domestic therapies cannot succeed any more than shock can "heal" infantile trauma or a warm puppy can restore the victim of Alzheimer's disease.[28] In his brilliant book *Masks, Transformation and Paradox*, A. David Napier describes the primal function of the mask as affirming ambiguity.[29] It is a marvelous, mitigating device of perceptual and philosophical meaning—not in the reduction of diversity but in its acceptance and in the principle of transformation. The basic mask is that of an animal, worn or carried in a ceremony or dance. Its message is that we are each two or more things at once—human and animal—an insight welcomed by out intuition. The mask and its dance remind us that transformation is closer to the heart of life than stasis or an abstract essence, that flesh and appearance mean more to our identity than ideology, that incarnation, not ideas or heaven, is what life and death are all about. The animal mask is a premier sign of our connection with animals as the framework of our humanity because it affirms the continuity arising in transience between forms.

* * *

Animals and their representations constitute essential elements in human mental life: cognition and psychogenesis, individuation, personal and social identity, surrogate and symbolic figuration, and the conscious and unconscious iconic repertoire by which emotion and other internal states are integrated, coded, and communicated. Animals connote fields of action and power—the objects of attention acquired during the evolution of human ecology as the neurophysiological structure of knowledge and speech. The characteristic mode of most of these processes is metaphorical. They came into existence in connection with a wild fauna directly experienced, typified in art, and translated in their metaphorical implications.

The substitution of a limited number of genetically deformed and phenotypically confusing species for the wild fauna may, through impaired perception, degrade the human capacity for self-knowledge. The loss of metaphorical distance between ourselves and wild animals and the incorporation of domestic animals as slaves in human society alter ourselves and our cosmos. Without distance and difference, the Others remain monsters of a terrifying jungle or, dissolved in our own unconscious minds, monsters of a chaotic and undifferentiated self.

NOTES

Acknowledgment: Thanks to Flo Krall for her critical reading and suggestions.

1. Claude Lévi-Struass, *The Savage Mind* (Chicago: University of Chicago Press, 1966), 115.

2. This is not to deny that many tribal peoples keep captive wild animals. The distinction, however, is between them and animals bred and genetically altered in captivity.

3. Bertram Lewin, *The Image and the Past* (New York: I.U.P., 1968).

4. Harry Jerison, *Brain Size and the Evolution of Mind* (New York: American Museum of Natural History, 1991).

5. E. H. Lenneberg, *Biological Foundations of Language* (New York: Wiley, 1967).

6. Eleanor Rosch, "Principles of Categorization," in Eleanor Rosch and Barbara B. Lloyd, eds., *Cognition and Categorization* (New York: Wiley, 1978).

7. James Fernandez, "Persuasions and Performances: The Beast in Every Body and the Metaphors of Everyman," *Daedalus* 101(1) (1972).

8. Only since the nineteenth century have we recognized the fairy tale as a distinct genre. Perhaps it crystallized out of the larger narrative domain in which it had been embedded, possibly because of the needs of the time. The idea that animals in fairy tales are one's own organic substrate comes from Bruno Bettelheim, *The Uses of Enchantment* (New York: Knopf, 1976).

9. Eligio Steven Gallegos, *Animals of the Four Windows* (Sante Fe: Moon Bear Press, 1991).

10. One thinks, for example, of the famous studies of the Nuer (most notably E. E. Evans-Pritchard's, *The Nuer: A Description of the Modes of Livelihood and Political Institutions of a Nilotic People*, Oxford, 1940), who create a conceptual cosmos from the cow. In general we tend to admire any culture that is traditional and intact and coherent. But the cow, in its earthly overabundance, devastates the environment, precipitating or confirming a rigid philosophy of duality. Among the Nuer the cow is the One. But as in all monotheisms, the cosmos as cow cannot be realized because the One inevitably evokes the Other—that which is not the cow. It is popular in academia to tell freshmen that polytheism does not exist because most peoples believe in a creator. But in real life the contrary is true: monotheism does not exist because the One inevitably implies the Other—the devil, the nonbelievers, and so forth. Perhaps the yearning for this impossible concept of unity is at the root of the modern hatred of nature—as our actual experience, to the contrary, as Lévi-Strauss says, is of "the ultimate discontinuity of reality."

11. Daniel E. Koshland, Jr., editorial, *Science* 244:1233 (June 16, 1989).

12. H. Spurway, "The Causes of Domestication: An Attempt to Integrate Some Ideas of Konrad Lorenz and Evolutionary Theory," *Journal of Genetics* 53: 325 (1955).

13. Konrad Lorenz, *Studies in Animal and Human Behavior*, vol. 2 (Cambridge: Harvard University Press, 1971).

14. In totemic society much time is spent watching wild animals, whose relationships include the normal violence of predation. Flo Krall has pointed out that this observation sublimates their own violent impulses. This is quite different from the intraspecies violence on television which, as a literal reference, is an incitement and model. Having nothing but the mild domestic animals to watch, our murderous species may keep its aggression repressed only so long before people start killing each other. In any case, as I have noted in *The Tender Carnivore and the Sacred Game*, subsistence peoples on small islands, without large mammals to hunt, regularize and institutionalize war.

15. Julia Kristeva, *Tales of Love* (New York: Columbia University Press, 1987), 119.

16. Ibid., 376.

17. Ibid., 212.

18. Ibid., 376.

19. Ibid., 120.

20. Ibid., 157.

21. Ibid., 159.

22. Ibid., 377.

23. Ibid., 120.

24. Ibid., 278.

25. William K. Gregory, *Our Face from Fish to Man* (New York: Capricorn, 1965).

26. This idea was suggested to me by Flo Krall (pers. comm.). See also Jane Flax, *Thinking Fragments* (Berkeley: University of California Press, 1990), and Harold Searles, *Counter-Transference and Related Subjects: Selected Papers* (New York: I.U.P., 1979).

27. Lévi-Strauss, *The Savage Mind*, 249.

28. Ambiguity marks the limits of taxonomy to deal with heterogeneity, especially in terms of growth, change, transformation, or metamorphosis. Inevitably there are creatures along the boundaries of definition. This is why all marginal forms are cognitively emphatic and all cultures give special attention to them and create composite, imaginary forms.

29. A. David Napier, *Masks, Transformation and Paradox* (Berkeley: University of California Press, 1986).

The Corvidean Millennium;
or, Letter from an Old Crow

I can now give you some information about the situation in my home-town of Galesburg since it was taken over by crows some years ago. The exact circumstances of the beginning are still hazy. Apparently the crows arrived with some preconceived plan and inveigled their way into the political system. Once they obtained a majority of the vote, they organized a bureaucracy for systematically converting the society into one like their own. I gather it was to be a kind of pilot project.

I was away at the time. However, I have been on friendly terms with crows elsewhere, and it was just this friendship that enabled me to see and copy the following letter:

"Dear Cousin,

"From the moment of our arrival the task of transforming them fell heavily upon us. But at last leisure is reappearing and it is possible to report to you on the events of the past few years. Fortunately, we are relentless by nature—thanks to centuries of harassing owls—and ulti-mate success is inevitable, though fraught with fantastic problems and danger. We began where individual habits are formed: in the personal and family life. As you know, humans have the disadvantage of never having lived in an egg. It surprised some of us in the beginning to find that, with all the helter-skelter of their lives, during which the unborn is carried internally here and there and everywhere, it is willing to come out at all.

"Luckily, we found some common ground in related matters, such as the keeping of a household. Although they had transferred to artificial caves, they do have an ancient tradition of nest-building by virtue of their primate connections. It was you, I believe, who once called to my attention the good sense displayed by the chimpanzees and gorillas in continuing this practice in a fashion more nearly like us.

"The very young are naked as our own are at first, but one waits in vain for the first handsome feather tracts to appear. Instead they develop rows of teeth, enough like ghostly grains of corn to make a sensitive crow shudder. Unfortunately, it will take many generations of selective breeding to rectify this weakness in their mammalian heritage.

"The young call instinctively for food as do our young. They are great imitators and learners of words. This trait is so common in them that one of the first questions we often ask about a young human is, 'Does it talk?' It usually does, becoming progressively verbose all its life, mimicking phrases it has heard and gabbing at a great rate about matters on which it is ignorant.

"It was through the main Subcommittee on the Improvement of Communication (of which I was cochair for the first three years) that we undertook to improve the situation. Naïvely we attempted at first to show the superiority of our own system. We outlined the repertoire of some thirty-odd instinctive calls, listing opposite each its communications function and the particular conditions, both physiological and external, that released it. My associates expected that they would see the advantage of an explicit and irrevocable meaning for each phrase—the wrangling, misunderstanding, and confusion it would put to rest—and the beauty of an inborn system of sounds for all the major communications.

"When this failed we attempted to excite human interest by alluding to the habit we share with them of sound mimicry. This was to gently elucidate the biological value of mimicking, provided it is not allowed to become a verbal epidemic. We explained the role of mimicry in developing local dialects, so that individuals may recognize their own subgroup at a distance and be spared the exposure of a social error. A special project was established by local civic clubs for training their children in our signals. But as the range of sound frequencies produced by crows was greater than their range, their young were handicapped. Then a pioneering experimental operation at the local research hospital crowned the project with success by showing that splitting their tongues greatly increased the number of Corvidean consonants their young could say.

"Further studies revealed some remarkable parallels to our own postjuvenile adaptations of infantile behavior. I am referring to our modified food-begging postures and calls that are used in courtship, and the feeding of one adult by another to strengthen the bond between them. They have a similar sort of ritualized activity in which, during courtship, baby talk is used and the members of the couple frequently pop tidbits into each other's mouths. We also found the familiar practice of mutual preening. These researches in the social adaptation of certain infantile and adult cooperative patterns broadened our understanding and our appreciation of them and hastened integration.

"Then, when integration seemed imminent, the famous 'women's riots' broke out. Flushed with our success with the language, we had pressed for improvements in the family structure. There were basic similarities on which we hoped to build a rapport; for instance, like us, they were basically monogamous, and the young lounge about the household watching the general nesting procedure for a year or two before striking off on their own. It was territoriality and seasonal flocking and migration that provoked street demonstrations.

"Looking back on it now, I think we failed to appreciate their ignorance of the insidious effects of crowding, so that when we proposed our minimal standard of 160 acres for each household their women reacted out of fear, excited by their mass culture, of being left alone. It was astonishing that a species related to the orangutan and the mountain lion could have degenerated to such a point. Our kind, I am pleased to say, has not forgotten since the mid-Cenozoic the necessity of territorial distribution of the population over the available space, limiting the population by forcing the supernumeraries to remain unmated, to wander about on the fringes, where the environment neutralizes them or where they are available as replacements to pop into vacancies. Had it not been for our coming I do not see how the Galesburgians could have avoided a disastrous population explosion.

"If their females were frightened of being alone, our solution through winter flocking seemed no more palatable. Indeed, the idea of abandoning the household for ten months of the year seemed to threaten the root of their family life. Our public relations people did a magnificent job of selling what was called the 'ten-month vacation,' the sloughing of the drab domestic routine and leaving the lonely isolation of the territory. They described in rich, slangy language the fun of aerial capering in a large flock, the excitement and mixing at the flock roosts, and the joys of wandering south for the winter. Many of the younger Galesburgians welcomed this new system but were checked by appeals from conservatives, asking them to keep tradition and the integrity of the home.

"One of our psychologists explained to me, however, that the conservatives' real motive was their reluctance to abandon year-round sexual activity. A secret delegation of them actually appealed to our Central

Council to incorporate this into our own way. But you can imagine the chaos that would follow, with thousands of sexually active individuals trying to fly south together. The waste in eggs alone, scattered from here to Texas, would impair our physical resources. I suppose that we should not have expected a very high level of social sophistication from a group enslaved by their instincts to perpetual pairing, with no time for developing higher levels of social intercourse.

"We were eventually forced to terminate the educational approach to this problem—as a combination of abandoning both menses and permanent home would certainly have released the mob violence so near the surface during those tense days. As you probably know, the problem was cleverly solved surgically by a simple operation on their pituitary glands. We avoided any mention of the gonads proper—which would have released all their superstitious fury—and by a slight poking about in their brains sensitized them, as we are, to minute changes in day length. Of course there are corresponding glandular changes. It is heartwarming to see them now, as the days lengthen into fall, restlessly scanning the skies, delicately responsive to the daily photoperiod.

"This brings me to that unfortunate Sturm und Drang period in the history of the Galesburgians—which I would avoid except that I know you are interested in administrative problems. It demonstrates beautifully what may happen when one part of a program is not precisely coordinated with another. During the fortieth year when some changes in personnel were being made (it had nothing to do with my department, of course), there was an awkward interstage between their old symbolic language and the adoption of our vocal signals—both the fixed kind and those for which the object is imprinted. It seems that the Posture Group was running behind the Sound-Signal Group. The Natives had lost most of their own language and had learned a great many sounds without the proper display that must accompany them. It was very frustrating for everyone, as they were nearly cut off without any communication. Also, the Commission on Proper Dress had taken away their clothes and issued them sensible garb like our own. Certain modifications in their physiques had also taken place. The result of all this was that it became impossible for them to recognize the sex of a stranger. None of them knew the subtle postures which identify males and females among us at a first meeting.

"This failure to recognize the sex of an attractive partner created a new problem for their morality. In our own broad-minded society no one gives a thought to homosexual pairing. Although it is not common when the sex-recognition signal has been properly made, it does happen sometimes, as you well know. The pair goes through the sequences of ritual, culminating in the construction of a nest, and then enjoys a great amount of good-natured give-and-take as they stand around, waiting for each other to lay the eggs. The humans, however, consider this a shocking arrangement. You can imagine the consternation for them when there was no visible means of distinguishing males from females and the pandemonium in which they were trying all the vocal signals they knew but getting them with the wrong postures. In the midst of this I saw a Galesburgian bowing in invitation to mutual preening and at the same time issuing a vocal warning that threatened the life of anyone who approached. I hope you will keep this fiasco in mind in case you should ever assume responsibility for an administrative complex.

"Once we had unsnarled this tangle we were ready to tackle other difficult details of social life. Not having an economic stratification in our own society, it was difficult for us at first to comprehend the problem. Also, they had political terms of abuse for what they thought our proposed system to be: *fascist, authoritarian, autocratic.* These of course were derived from mutations of their own system, which had only superficial similarity to ours. Unlike either the fascist or economic democratic state, we described our society as combining control and flexibility. No better scheme has ever been devised than our dominance-hierarchy system for providing the individual with exact information on his place and at the same time making it possible for him to move up or down.

"When we first built models of this system by incorporating it into their school classrooms, it was popularly referred to as a *peck order.* But we recognize peck order as a primitive social design of the crude sort enjoyed by chickens. It is true that there is a top dominant individual in both systems. But the silly fowl have a simple linear sort of arrangement with a single set of criteria operating throughout. One of our most difficult tasks was to convince them of the elaborateness possible in our dominance-hierarchy order.

"We pointed out that, like those superior mammals the brown rats, we have added some elegant refinements to dominance order in feeding.

Bottom-level birds seldom if ever go to the food source, such as a carcass. This reduces even further the chance of congestion and antagonism. Instead they behave as though apprenticed to high-ranking birds (as all of us do as yearlings), diligently but respectfully following them on the ground or watching them from a perch. The ranking individuals do not abandon the carcass after eating but continue to carry off mouthfuls, which they poke into clumps of grass or under things. Then as they return for more a watcher rushes over, plucks the food out, and eats it. What better example of noblesse oblige could there be in a hierarchical order?

"Another successful move was to demonstrate how our system made it possible for intellectuals and artists to rise as high as the middle of the society without being subject to actual pecking by anyone except each other. We appealed to their females by describing the ease with which a female can marry into a social level without having to move up by stages. We showed that a neurotic in the group can find a tolerable existence without descending to doormat status. Once we had these minority groups on our side we set forth our principal argument—which, as we expected, was welcomed in the press, pulpits, and political arena. This, of course, was our ritualization of fighting in routine situations, the advantage of symbolic struggle to avoid disablement. Once the Posture Group had done its work, the Natives were prepared to try ritualized fighting in their daily social conflicts, in substitution for fistfighting as well as for their barbaric practices of conniving, snobbery, character assassination, economic stratification, and segregation. We were delighted to find even faded behaviors, suggesting that they had ritualized conflict at some time in the distant past, as seen in the submission posture of bowing and doffing the hat, in various forms of games, and in dancing and other arts. Intimidation by bluff and threat can be beautiful to watch, as you well know, dear Cousin, when little physical damage is likely. Naturally some of them were reluctant, but they at least had the consolation of knowing that during the nesting season each would always win any conflict in his own territory. This fact, which is the cornerstone of our territorial distribution, seemed to appeal to their sense of democracy.

"As you see, only when we had obtained public support did we exhibit the sterner details. By this time several groups, such as the Young Democrats and local unions, had organized on our behalf and were joined or not opposed by other groups—except, of course, the D.A.R. and the American Legion. Among the more delicate matters to be broached was the absence of the concept of 'crime' in our society. You might at first suppose that they would have been delighted to form a society without crime, but what this meant to them was that nothing was unlawful. We patiently explained that this is because all behaviors are responses to certain signals in the environment and are inherently patterned—or else are random searching activities for such signals. The "responsibility" of the individual in this system is, for example, to steal food from another who is vulnerable to him in the social order rather than from one who will successfully defend his possession of it. In case of error, the punishment is failure and humiliation; there is no further penalty against the culprit. They had some difficulty understanding how one's responsibility to the society was to steal, in spite of the beautiful simplicity of this arrangement.

"They were inclined to think that this permitted aggression against the 'weaker' individuals, violated 'property rights,' and was in fact 'unethical.' It was necessary to reply that our hierarchy provided a secure place for what they called 'weaker' individuals, but not by pretending that everyone was identical; that in our view the world was not a bundle of 'property rights' but that we instead were guests in it, not its owners; and that the greatest morality consisted in the total acceptance of ourselves, socially as well as personally.

"I suppose they were especially interested in the matter of stealing because they are so strongly subject to its lure. One of the first characteristics we noted on arrival was their hypnotic fascination with brightly colored objects. Their offspring particularly were always picking them up and carrying them about, often secreting them in some corner or in a pocket.

"Aware that we were working toward the sensitive areas of deepest feelings, we carefully outlined two more important matters in the Corvidean society: the recruitment of new flock members and the killing of some. I was not referring to the coming together of family groups to form the winter flocks but the addition to it of wandering individuals. Among humans the arrival of a stranger into a locality was sometimes a distressing sight: he was completely ignored, ostracized because of his annual earning, insulted by a commercial welcoming agency, or harassed because of a creed or racial trait. In a number of cleverly edited pamphlets we showed our contrasting way, where only the dominant individual pecks the newcomer intensely upon his arrival. I mean, of course, young birds less than two years old, whose mouths have not yet turned from red to black inside. As the blows fall, the youngster adopts a food-begging posture and his red palate inhibits further aggression. It is an irresistible signal to number-one to desist. It is typical of our social efficiency to transform juvenile food-begging behavior into an adolescent function of appeasement, and still later, as I already mentioned, into a courtship stance.

"This brief treatment is greatly to the benefit of the newcomer. The period of recuperation is sufficient for him to observe the flock interrelationships. The shock of his slight wounds deters him from feeding, keeping him out of the main focus of further conflict. After a time of appraisal and recovery, he chooses his own level at which to attempt to enter the flock and initiates the move himself.

"I know you are wondering, Cousin, how we introduced the procedure of killing flock doormats. Let me assure you that we picked our best brains for the task. We approached the subject by emphasizing that this was only done as a necessity, on occasions when the flock had grown too large for its supporting resources. Since our hierarchy is a web of relationships, a pulse of drastic realignments follows the loss of a member from the center or head of the flock. This is bad, as everyone is nervous and upset for many days when it happens. To avoid this disorganization and to remove the poorest combination of chromosomes from the flock, the dominant member selects the victim from the dregs of the society and destroys him. The bottom of the hierarchy is occupied by the most poorly integrated individuals. If number-one is capable and experienced—which he must be to hold his position—the act is swift and sure, with powerful jabs aimed at the brain. The body, especially the head, may then be eaten by members of the flock who are in need of protein. But there are always, as you know, Cousin, some of us who prefer not to eat crow under any circumstances.

"Although I repeatedly emphasized the small number of occasions on which this drastic action was necessary—as overpopulation was normally prevented by territorial nesting—there was some opposition. They did not object to the death sentence but to cannibalism. It was inevitable that this should lead to the whole subject of death, exposing one of their most barbaric superstitions and fears. Even as I write this, two of our historians are preparing the final draft of a monograph entitled *A Case of Ecological Schizophrenia: The Artificial Cleavage between a Species and Its Biological Heritage* (to be published by Nevermore Press next year). Their pathological anxiety about death and preservation of the body, thus keeping it out of the normal food and energy circuits, has no parallel in the animal kingdom. It was difficult at this point in our discussions with the humans to keep our tempers. So long as they remained a minor species sparsely distributed, as they were for more than 500 thousand years, what they did with their dead did not much affect the equilibrium of the landscape. What incensed our committee was that they had devastated so many other species and altered so drastically the earth's surface to divert more nutrient element flow into human protoplasm that it seemed the worst sort of snobbery and short-sightedness for them to block it at that point. Our Natural Resources Committee was especially upset about tying up that much phosphorous and calcium.

"The humans seemed to have no sense of their responsibility except to their own species. They were unaware of the patterns of mutual relationship shared by all organisms. We Corvidae have always been extremely proud of the elegant food web to which we belong, and we cherish the various food chains in it with all the fervor they lavish on exclusiveness in their social clubs. Take, for instance, my own special interest: the aster leaf–caterpillar–beetle–white-footed mouse–crow–Cooper's hawk food chain. To belong to such a unique and ancient association is to me a source of great pride. It is unimaginable that our species should insist that the chain be broken before it reaches the Cooper's hawk, or, indeed, before the bacteria render the nutrients of the Cooper's hawk once again available to the aster plant. It has always given me a special sense of pleasure to know that we belong to a greater variety of such food chains than any other birds. When you include other members of the family Corvidae—the ravens, magpies, nutcrackers, and jays—you find that we have much to be proud of in the diversity of our ecological attachments.

"Coming back to the aster leaf–Cooper's hawk food chain, another human who was noted for his tenderheartedness declared that he was willing to become food for a hawk, or even for one of his fellows, but that he could not bring himself to kill the handsome little white-footed mouse. He foolishly supposed that we were inviting them to join the aster leaf–Cooper's hawk food chain, as though one joined such a chain by being voted in and taking an initiation!

"Aside from that naïve mistake, his comment exposed the most staggering cleavage between them and the natural world. For all his repugnance for killing, he represented the worst kind of killer in the history of life. He supposed that by delegating the killing he was not longer responsible. The spinach plants and clams and poultry and seeds of corn that he consumed without the slightest reflection on the toll levied to fill his plate were indeed not killed by him. They were killed by professionals and the extensions of themselves through machines.

"They had taken in their society the lamb as the symbol of innocence and gentleness. To eat such a creature one should know it very well and love it greatly—as we do the mice; live with it in the same habitat in all the seasons; then kill it with one's mind and heart full of nothing else, in a joy so deep that no words can tell it. Instead, their lambs were driven through a series of degradations and filth to a mean death while they prated about the ethics of killing and mindlessly munched their lamb chops.

"Only the hunter and the gardener know directly, dear Cousin, the ritual value of eating and do not attempt to split experience into spiritual and fleshy fragments. And I need not remind you that we are both hunters and gardeners. I myself have planted many a seed by poking it into a crevice or burying it. Indeed, one might see in our frugal habit of hiding what we cannot eat an age-old horticultural enterprise. This interest in plants keeps our carnivorousness in perspective, I think. Otherwise we would become rigid and narrow like the hawks.

"In our long discussions on these matters we found that they had little conception of the responsibility of the hunter to the prey or the eater to the eaten. To explain this, I told them the remarkable story of the relationship of crows and ducks in the prairie marshes of the North. (Perhaps you do not remember the circumstances, Cousin; no reflection on your broad knowledge is intended, of course; I know you have been very busy at home with the Committee on Integration to break down the color bar, to end prejudice against albinos and admit them to equal footing. Please remind the Committee of my research showing that differences in feather pigmentation between black and white in no way reflect any inferiority in the latter.)

"But to get back to the ducks. The poor silly things all begin to nest the same week and have, as you know, insisted ever since the Eocene on putting the eggs in a little hump of grass an inch or so above the water level. This is convenient for the young, but storms occurring during the week of hatching kill a high proportion, as young ducklings are very susceptible to drenching and exposure. By the time the ducklings have hatched it is too late in the event of such a fatal deluge to start another nest. That a chance storm does not decimate the whole year's hatch at this vulnerable age is to the credit of the local crows. From the day the first duck eggs are laid the area is searched over by crows, as the food value of duck eggs adds a lovely bloom to our black feathers. Whenever a duck nest is broken up and the eggs destroyed during the incubation period, the ducks go off and start another. As a consequence, the hatch of ducklings in the marsh is spread over several weeks and no single storm hurts more than a few of them. Were it not for the crows, the ducks could lose most of their young as often as every other year.

"Now the humans simply had no place in their barnyard philosophy for this kind of obligation, in which the crows eat unhatched ducks. To them wild animals were, instead, good or bad, depending on whether they rank higher or lower than man on the food chains. I say barnyard because their system of values was derived from the rural culture of Europe, in which a "bad" animal was one eating the farmer's chicken, tomato, or dog, be it snake, worm, or flea. It was then condemned as a species. Animals neither in the domestic barnyard nor competing with people to eat the stupid barnyard creatures were in yet a third category labeled "useless." I ask you, Cousin, how does one deal with such an absurdity? Since nature is a continuous fabric, how can one part of it be useless? It is like saying that the inside of a coat is useless because it cannot be seen!

"As you can imagine, they were slow adopting this broader view in their ethics. The ancient Greeks, from whom they derived the idea of an ethic and who shaped the modern human mind, had a religion that compensated somewhat for treating only human intercourse in that ethic. To them all beings were sentient or had anthropomorphic astrological significance. We should feel fortunate that we are making any progress at all in view of the economic-commercial schema to which this homocentric ethic has been geared in a technological society.

"If I seem somewhat discouraged, it is only because a temporary difficulty has me somewhat upset. I really do think we are making excellent progress on the bigger problems. I should not bother you with a minor matter, but you might find it instructive. We had finally convinced the Galesburgians that nests were better than houses and had begun a ten-year antihousing program to get them out of houses and into the trees. We had no sooner gotten under way with the support of the mayor and Council than a disease struck the elm trees. Overnight there developed an acute nesting-site shortage. This will delay the transition somewhat, but it does have its bright side. It was possible for us to retain the best remaining nesting sites for some of the socially prominent families in the city. Subsequently, tree nesting became a status symbol, and there is a growing public clamor for an intensified tree-planting program. There have even been some attempts at nesting by the Galesburgians in television aerials—that's rather silly, don't you think?

"Well, Cousin, that is where matters stand in our democratic transformation of their culture toward something nearer the heart of the majority. We hope to give wings, so to speak, to their pedestrian society. Which reminds me, I am flying west tomorrow to examine the village life and mythology of the Crow Indians, a subtribe of the Sioux in the Yellowstone Country. Friends here suggested that I detour to see various places of interest along the way, but of course I had to explain that that was impossible—we always fly direct.

"Your affectionate and faithful cousin, James."

Place and Human Development

There are two approaches to the study of place, expressions of a larger split that penetrates all modern scholarship and makes a dyad of public opinion. The majority position is that the pronoun is largely self-formulated, that uniqueness is grasped and identity consciously shaped, or at least chosen from the anomalous streams of circumstance—economic, social, and political. Current writing is chockablock with this view of the autonomy of place and self. It rests on the assumption that the earth, like the body, is a nascent expanse of no-place until a kind of cartography has posted it. Its ideal is to escape the deterministic aspects in favor of pure creativity, what Sibyl Moholy-Nagy called the modern "cult of eccentric originality," in which architectural genius builds unprecedented forms in otherwise mundane terrain, or, I suppose in the case of the self, it follows the Renaissance dictum that one's life should be a work of art.[1]

The minority and more interesting position is that there are either compelling local forces in the earth itself, be they telluric, chthonic, or autochthonous, creating peculiar fields of energy to which people are more or less sensitive and which are typically discovered rather than constructed. Obversely, there are intrinsic, structured, chronological analogues in the individual human being; identity coming from within, that is, instead of superimposed or fabricated. History and the ideological mind tend to designate locations, as when we place a plaque upon a site or declare a society Marxist or a trophic type vegetarian. The contrary view is prior to history and has been generally accorded a grudging place in a world of relativistic and existential values. This constraining *given* is shaped by the natural history of the species.

In 1973, Mayer Spivak called attention to a facet of animal ethology often unnoticed—that large, mobile mammals use many different habitats at particular times of the season or life cycle.[2] To be ecologically nichebound, as all living things are, is not always to be habitat restricted. This does not imply, except to a kind of credulous hubris, a liberation from natural context but, on the contrary, a commitment to sequential diversity.

It is an appropriate observation from which to consider human ontogeny—the development of the individual: the neotenous estate, the retardation of development and extended immaturity—biological specializations of a high order. Misunderstood in the humanistic literature as a kind of freedom to erect an autotelic self-world, in a cultural milieu that might be characterized as autistic, by my dictionary "a form of schizophrenia typified by acting out," the "freedom" of childhood has been misunderstood. No wonder there has been historical disagreement as to whether there even was a childhood in the past.

To return to ontogeny, it is, succinctly, a program, genetically coded, whose successive phases are keyed to environmental and social signals. The nonhuman coplayers in this maturational drama or game include an array of animals, the perception and identification of which provide both the cognitive instruments of categorical intelligence and projected fragments of the inchoate self in their respective stereotyped behaviors, spread across a landscape. As a Siberian myth tells it, all the body organs once lived independently as creatures. Something of the sort also keys the child's emotional states, feelings, and behaviors to the concrete forms of nature, an external reality swallowed by introjection and internalized to make a self. It is a kind of shadow play of the sacramental rite of eating the god's body to attain grace. The animal infinitives—to duck, to bear, to grouse, to bug—constitute a mosaic of the living repertoire gathered in a finely tuned, unconscious, mimetic, critical-period attention, speech, and nurturance.

Ecologically, animals are inseparable from place, except as we see them domestically blunted, insane in zoos, or petted as meek surrogates. Their reality construes a field or ground. Now the possibilities open wide for exploring this framework of habitat of the animals, themselves perceived as dispersed elements of a self, to whose gathering (or hunting) the ontogenetic program is inexorably committed.

Beyond the animals proper there are two more explicit processes in which a copula occurs in growth, between the individual and a *genius loci*. Perhaps the most remarkable document on childhood in this century is Edith Cobb's *Ecology of Imagination in Childhood*. Surveying the lives of geniuses, she noticed a common thread—the return in moments of creative meditation to the place of childhood in imagination or sometimes physically, a trip that helped toward a solution to a problem. The original meaning of the term *genius loci* referred to a unique sacred power. What was it, Cobb asked, about the original experience that made it useful to the psyche in a recapitulated travel across the juvenile home range; in what sense was it an organizing force?[3]

She concluded that the adult faith and intuition that order permeates the cosmos, that no bit of data or bizarre idea was truly disparate, that searching would be rewarded, extends from the singular imprint of an intensely inhabited space of about thirty-five acres at a crucial time of life. Played through, the child's transit, time and again, locked this literal, objective reality into an unforgettable screen, through which other, novel objects of the mind would be envisioned by the questing adult as though they were details of a landscape. Just as the mnemonist studied for thirty years by A. R. Luria "placed" images for later retrieval along a path in the mind's eye, at some less conscious level a holding ground is absorbed.[4] The juvenile home range is a tiny universe, whose trees, rabbits, culverts, and fences probably register some kind of metaphorical series whose branching, skittery fleetness, subterranean connecting, and boundary-marking functions in relation to a speculative field of half-formed and elusive ideas follows a paradigmatic system of relationships. An anatomical model for this unlikely neural representation of place is seen in the fundus of the eyes of vertebrates, where the colored oil droplets in the cells of the retina, differing according to the frequencies of light in different parts of the visual field, form an eerie landscape that can be seen with an ophthalmoscope. Edith Cobb's own genius has given us insight into the primordial meaning of coherence as a function of a specific, tangible, ecology, swallowed by the nine-year-old in repeated excursions.

The means and implications of this order-making process remain to be understood. We have yet to learn to what ultimate conjugations of sacred forms, symbols, avatars, and ritual acts the streams, tree houses, pathways, and mean dogs lead, for they are the fellow players in moments of acute awareness who occupy the space labeled by one of my students, "the chipped-tooth, dead-cat landscape." Does pretending to be a machine instead of a mounted horseman make any difference? What counterpart is possible in the city's fabric to the flowing lines of the countryside or the Otherness of the wild places?

As for the eventual sacred forms and symbols, their link to a natural matrix is enhanced by another series of events, which constitute the second of these place-bound episodes in ontogeny. This is the setting of adolescent transformation, a culturally mediated passage whose understructure is puberty. Its religious connotations have been dealt with at length by Micea Eliade and others, and in another plane, its psycho-physical signs and pathologies constitute a whole subfield of psychiatry. That the pubertal child has a precipitate mentality and emotional and intellectual prevision of unknown events and trials is a culminating phenomenon of childhood. Its focus is an initiation into adult status, but the complex includes separation from family, instruction by elders, tests of endurance and pain, trials of solitude, vision, dream, and rituals of rebirth. There is reason to think that modern anomie and disaffection may have roots in the failure of society to respond to this complement of time-critical needs.

Central to the imagery of the adolescent passage is the creation of an ideal world beyond the two already experienced as ontal and phenomeno-logical. It is a transition in the imaginative faculty from a literal to a figura-tive place, from familiar temporal reality to the dreamtime. One of its instruments is, to use the psychiatric jargon, the *transitional object*. Unlike that of the small child, the doll, or blanket, it now refers to places and objects of special significance, stories told, and the enacted metaphors of rit-uals of death and rebirth in which a leap is made from the natural world to the cosmos.

The transitional objects of adolescence belong to the category of *objets trouvés*, which are completed in art. The most enduring set of these in human history is probably the petroglyphs and pictographs of the Old World, paleolithic caves. The primodial setting is the cave, and the mime of rebirthing is usually preceded by physical travels in a terrain of hills, springs, and special trees now given a second meaning in myth by the poetics of fire and the esoteric forms of combinational animal images. I said a "leap" from one world to the other, but that term is not exactly right, even though dancing is important to it. Eliade may have been wrong in postu-lating the exclusive distinction between sacred and profane worlds, an opposition that fits mainly peasant Christian conceptions or other Indo-Aryan dualisms. The older function is not a leap from this world to the other but a leap like the shuttle in a loom, binding and informing rather than departing. This world for the peasant tends to be a meager model for the hereafter. More typically, the ceremonial enactment obtains identity with a mythical past not by transcending the physical reality, but by enlarging its meaning and obliterating the kind of time scheme that char-acterizes historical societies.

The typical adolescent preoccupation with her or his own body is, from this persecutive, crucial, for the introjection of space that constructs the body in the child ceases to be the palpable place it was. The body is to be reconstructed along with the world. The special places of adolescent initiation therefore embody a story, simultaneously altering the individual's interior and the whole landscape. The concept of place is first a world of concrete, named things and relations, the tacit equivalents of an inner constellation, imprinted as the first model of order. Merely literal place then metamorphoses in adolescence, becoming a syntax for the description of an ideal world.

* * *

I have mentioned briefly three moments of correspondence between inner and outer worlds, in which specific features of the earth and its fauna provide the child with sensory figures corresponding to latent elements of the self. Body and space are construed in tandem. A closure occurs: specific places and creatures become unique to the self, obtained through a temporal window.

The juvenile terrain and fauna, and the adolescent map of environmental apotheosis, are bondings. The first refer to a concrete matrix encountered in the protorituals of play. The second is a triple matrix of spatial counterparts—the body, the terrain, and the cosmos.

Two associated issues can be only noted here. The first is freedom. In the world of identity by superposing (the hierarchy of the corporation) or by self-proclaimed ideology (communism, libertarianism, feminism)—that is, the entrepreneurial world of making place by free seizure—we come finally and paradoxically to live alienated lives, fostered by myths celebrating the arrogance of Jonathan Livingston Seagull and the power of strip-mining. But seagulls and the cycle of buried carbon can also sustain other myths—those of keeping, of freedom as defined by Rollo May as the affirmation of imitations.[5] We are free, culturally speaking, to internalize the bird and the terrain either way we wish, but we are probably not free to control the consequences.

The second point involves the question, what happens to the child who misses Nature? In a cement or desertified home range without creatures, what is the ontogenetic outcome? If the adolescent cosmosizing process follows a barren childhood encounter, I presume it proceeds to invent forms without an organic substrate of Otherness or the coherence of dense uniqueness, that is, to project infantile concerns on the empty stage of the world, and to cosmosize that placelessness as a battleground in which fantastic ideologies struggle for power in the mode of juvenile heroics.

Such an ideal world would lack both the toleration of Otherness in its animal forms outside and inside us and the ecological fabric giving meaning to space and time. Madness is the kindest word for such a world.

In terms of American experience some of this conceptualizing of the constituents of an identity takes on distinct qualities. Among these are the geographic mobility of families, the wide density of experience of animal diversity, and the various efforts to meet adolescent needs. Americans have among them surviving Indian groups among whom the psycho-ecological needs of the child are reasonably well met—some groups of which are increasingly available to Caucasian Americans in a didactic way. Americans are paradoxically closer to and farther from indigenous fauna and landscapes than their European cousins. Their lack of traditional identifying events that conserve socially shaping activities in Europe make their deracinated condition and yet, in a way, impel the society to explore its means and perhaps to recover these Paleolithic modes of shaping which our Pleistocene bodies still "expect." Despite the rapacious work of the nineteenth-century lumber barons and the twentieth-century corporate energy moguls in homogenizing the continent, much wildness remains, and the fluidity of society is itself perhaps a doorway to the realization of Roxy Gordon's statement that "real revolution is born from genetic memories of ancient reality."

* * *

However arbitrary the details of play or the content of myths of origin, or the choreography of ceremony, they constitute the inescapable drama of the unfolding personality, bound to the necessity of a diverse and healthy surrounding. Maturity is strangely linked to the nature of hatching-places, its outcome a measure of the resonance of inner and outer landscapes.

NOTES

1. Sibyl Moholy-Nagy, "On the Environmental Brink," *Landscape* 17(3) (1968).
2. Mayer Spivak, "Archetypal Place," *Columbia University Forum* October 1973.
3. Edith Cobb, *The Ecology of Imagination in Childhood* (New York: Columbia University Press, 1977).
4. A. R. Luria, *The Mind of a Mnemonist* (New York: Basic Books, 1968).
5. Rollo May, *Love and Will* (New York: Norton, 1969).

Place in American Culture

Place in American life begins with one of the most distinctive bodies of art we have produced. A group of about forty painters, sometimes called the Hudson River school, produced several thousand canvases of New England landscapes in only about three decades before the Civil War. Almost everyone is familiar in a general way with these pictures, some of which are in the larger American galleries. Their real abundance is not often appreciated, for they are stored by the hundreds in gallery cellars and are owned privately in great numbers. The question of why they were painted has been seldom asked and poorly answered. Usually the subject is confined to the notion of frontiers, a sense of scale, conquering the wilderness, or it emerges within the history of art, a consideration of the styles, training, and influences of the painters in a cultural era called Romanticism.

In that idiom the paintings are provincial expressions of painterly traditions of England and northwestern Europe, with its centuries of "Italianate" influences and Dutch predecessors. But why they are set in such a restricted geography, compressed in time, homogeneous in subject and naturalism, cannot be explained simply as American manifestations of a larger tradition. The American paintings are unique in several ways, mainly in that they are portraits of places, especially wild places.

Their homogeneity disappears when they are examined through the eyes of the naturalist. What to the ordinary viewer appear to be repetitive scenes of forest and mountain are actually meticulous explorations of the diversity within a unity, the variations on themes of a certain biome and geography. Seldom, perhaps never, in modern Western art has there been such focused scrutiny of the appearance of a region, especially of its geological and botanical details.

If you drive through these landscapes today, you see a phenomenon that is happening almost everywhere—changes whose origins and materials, designs and purposes do not originate locally: roads, buildings, transmission lines, vehicles—the whole paraphernalia of technophilia—

brought from elsewhere, from common centers of planning and production. The countrysides produced by such means are themselves duplications of one another. Among an educated and somewhat supercilious class this produces a familiar lament: the loss of native crafts and local industry to the enveloping tentacles of centralized, political, and industrial forces in a society whose mastery of nature is based on the replication of machines and the modification of terrain to make them work.

But why do we mourn the loss of the native ambience? Critics say it is only self-indulgent nostalgia, sentimental attachment for a past when life was simpler or cleaner—largely a self-deluded view. This opinion is echoed by the minions of change: our attachment to obsolete, old-fashioned economies, skill, or styles is but a romantic dream. Those afraid of progress are said to cling to the past and impede growth and affluence. Indicted by this accusation of weakness and immaturity we yield to its logic, and the worldwide blenderizing goes on, with the blessing of the industrial state, converting the globe piece by piece into a business network of uniform parts and identical places.

The angry reaction to romanticism—past or present—has just about run its course. Whatever the foibles of romanticism, its concern for the organic, for the wholeness of things, for feeling as well as reason, has outlived the spurts of bad taste and emotional affectation that it generated. Even so, the romantic in each of us remains a whipping boy, an excuse for our own lack of conviction. The yearning for scenery, nostalgia for the family farm, respite from the urban roar, a sensitivity to wildlife: these are regarded still as the indulgences of those who can afford them, symbols of wealth and privilege. Enthusiasm for the local and unique, the customs and environments that differentiate one region from another, handicrafts and skills, things homemade, exotic places, "primitive" peoples—all are associated with the idle amusement of the few.

And, in a sense, they should be scorned. They have been the playthings of fossil-fuelman, a part of his connoisseurship of cute objects and colorful places. Parochial values are debauched by layers of exploitation and misunderstanding. Picturesqueness reduces everything it touches to mere surfaces. Ever since the eighteenth century the gentility went about Italy speaking of the *genius loci* as though it were a landscape painting. They pimped for tourism's whoring of place. The admirers of landscape were not the opponents but the agents of the Faustian spirit, the founders of the *aesthetic* passion or nature, who would facilitate the destruction of peoples and places for the arts by perceiving them as "local color" and subordinating them through commercial travel to the movers and changers at the heart of Progress.

To the old Romans, whose poetry created the pastoral, the *genius loci* had not meant picturesqueness at all but referred to a tutelar divinity, a guardian spirit. It had been the same among the Greeks, whose temples were expressions of the character of particular goddesses in whose laps the temples were placed. The subtlety with which they were accommodated to the terrain extended even to the configuration of the horizon; the temple passages were designed to guide ritual processions whose central themes were a dialogue between the people and the earth.[1]

However much they admired the old arts, educated gentlemen more than a millennium later could feel little of the old pagan interior sense in which these sacred places were experienced as part of themselves. The Jews and Christians had methodically sought out the old shrines and turned them into churches whose essential purpose was to direct the religious inspiration away from that place.[2] To bishops who consecrated them and the liturgy they followed referred to a Holy Land elsewhere, a heaven and hell which were nowhere and everywhere.

In his widely read book *The Sacred and the Profane*, Mircea Eliade has instructed a whole generation on the history of sacred places: the rites and ceremonies that "cosmosize" a hearth or an altar.[3] But there is for him no real chithonic, no real spirits, only human beliefs. Eliade is a Christian historian. Although at pains to insist on the religious man's loyalty to the heterogeneity of space, he sees it only as something made by men. The intrinsic qualities of the spring or cave or mountain are for him little more than markers. There is not the slightest hint that the spiritual entities that the Romans, Greeks, and scores of so-called primitives conceived as indwelling and indigenous were anything but cultural assertions. One always called in the forces from a centralized heaven the way one dials a long-distance operator.

Of course the Christians did not invent this making of place by will and designation. The ancient civilizations of the Near East are speckled with temples built where they would be convenient to the bureaucracy, the keepers of the grain and the army barracks. The shift of attention away from the uniqueness of habitat began long before the Church fathers declared that all places on this earth are pretty much the same. Eliade's view is ultimately no different from that of municipal street-namers: the world behind the human façade is homogeneous. One *founds* rather than *discovers* place. It doesn't matter whether a priest blesses an altar or the mayor cuts a ribbon. The autochthonous forces by which the Earth speaks are not part of nature but only elements in myths through which the peasants rationalize their designations of sacredness. It assumes that we know what is given and what is made.

But the polarity of the given and the made will not go away. It is the duality at the heart of knowledge, the central enigma of our private and collective identities. Arthur Modell has observed that the painting and sculpturing of the Paleolithic caves of southern Europe often use the erosional forms of the rock as a part or whole of the animal figure.[4] This "transitional" synthesis of what is there and what is created externalizes the linked polarity between the culture, the artist, and the stone, between our selves and our bodies. The artist formalizes that tension which is the core of the maturing self-consciousness.

Art, says Modell, is always a love affair with the world. Henry Moore, the sculptor, liberates his massive reclining figures from within the stone; they do not escape so much as articulate their own particular mineral substance. Like the cave art, his work is a search for self that was not solely defined by the acts of transcendence and domination that energized romantic tragedy, feelings for which the self-styled "neoclassicist" and modern highway–pipeline–parking lot builders cannot conceal their antipathy. Romantic tears seem to them like infantile weakness. The temples and caves were part of an irretrievable past, they said.

And infantile they were, for the stresses of deprivation have regressive effects. Miss a developmental episode that belongs in your sixth year, and in some respect you never get older. Even when the connections are made you are drawn back sometimes to your roots for renewal. Between the natural and human, given and made, the other and the self, what the romantics sought was "a place in which to discover a self." This apt phrase of Edith Cobb's is a way of describing a childhood process by which the terrain and its natural things become a model of cognitive structure for the plastic, order-seeking juvenile.[5]

The child, she says, seeks to make a world the way the world is made. Her studies of the biographies of geniuses led her to the conclusion that the terrain itself provided the durable gestalt upon which the intellect germinated. Home range for the eight-year-old is the prime, patterned, concrete reality in life, upon which the wavering and nubile powers of memory and logic cling and develop, like seals climbing out onto the rocks to give birth.

In his book *The Mind of a Mnemonist*, A. R. Luria describes a man whom he observed and studied for many years, a man whose phenomenal power of memory enabled him to remember everything and anything—all the words in all the paragraphs in all the pages he ever read—a man who could repeat the names of a hundred spices or a hundred flowers in order, regardless of how much time had elapsed since he saw the list.[6] To Luria's question "How do you do it?" the man replied that he could visually recall the book pages and that when he was given a list he took an imaginary walk in a landscape, placing each object in view along the path. To recall them he had only to picture that place again and his walk through it, to see the objects. The man was abnormal. Most of us have the blessing of remembering trivia only in the unconscious, if at all. But his anomaly was a clue to a strange and necessary relationship between place and mind.

Cobb's study of genius was a search for the genesis of thought and creativity by studying the lives of the gifted. That genius is both the "spirit of place" in its classical sense and a personal divine spark among the most powerful minds is not coincidental. That early formation of the self in children who are yet untouched by ideology is a growing awareness of one's own anatomy, the discreteness of body parts: both organs within and complexity of surface, sensory location, and feelings and moods as well. Experience for a small infant is a formless sea of feelings that engulf the individual. Their subjective separation requires a spatial detachment. They are intuited from external models and introjected. The constituents of self need externality and distance to be comprehended. Considering how admirable the human individual is, as Shakespeare tirelessly reminded us, the burden on the environment is great indeed. Diversity, richness, all those terms of multiplicity that describe a heterogeneous world, have been demonstrated repeatedly by biologists as essential to the development of intelligence. From nutritional and environmental studies of laboratory maze-running rats[7] to the observation of babies with and without playpens,[8] institutionalized children,[9] and the psychology of the playground,[10] the evidence is strong that heterogeneity is like an essential nutrient.

But how does it work? You cannot, after all, just put a baby in a bag with a thousand objects and shake well before using. Claude Lévi-Strauss believes that the species system of plants and animals is a durable,

dependable concrete model for the development of the powers of cate-
gorizing, or basic cognition. Edith Cobb holds that the fixation on terrain is an organizing process by which the precept of relatedness is inte-
riorized. White and his associates at Harvard find that the intelligence of
children emerges relative to a spatial movement among objects, coupled
with naming. All of these imply real changes in the nervous system. We
can visualize the possibilities at the individual level among people by an
analogy to differences that occur in the nervous systems of different
species of animals who live in different habitats.

A series of paintings done in about 1901 and published by the Royal
Society represent a kind of extension of the terrain into the inner eye.[11]
These were the first color representations of the ocular fundus, the
retinas of living animals. The patterns have a likeness to environments
that is inescapable among those animals living in horizontally structured
habitats—sheep, horses, lions, dogs, humans—while the fundus of
others, who live in the trees or underwater, lacks the landscapelike patterns. There is nothing mystical about this. The structure and content of
the visual cells correspond to differences in the frequencies of light in
sections of the visual field. In open terrestrial habitats the field is roughly
divided into bands of light from the sky, the ground, and the horizon. It
suggests in a way perhaps more symbolic than evidential the capacity for
structuring the nervous system on an external model. To me these retinas
look like impressionist landscapes. In an evolutionary sense, the habitat
has impressed its form upon the neural tissue, and the individual
organism seeks out those places in which its sensory and nervous systems
work, orienting its head and eyes to light patterns matching its visual
anatomy.

The individual's intrinsic needs all have spatial settings that are not
inventions but a mammalian heritage. In a seminal paper on what he calls
"archetypal place," Mayer Spivak has enumerated thirteen subdivisions
of the habitat.[12] One can live without special places for resting, feeding,
conviviality, grooming, courting, and so on, but without them we
become, like deprived, captive mammals, increasingly stressed and
pathological.[13] Such places are not arbitrarily labeled in this ancient tradition of mammals but are the psychological and physical prerequisites
of the different behaviors. All of them have perceptual and psychological
dimensions.

The development of this continuity of internal and external, the reciprocity of place and person, takes place at every level of experience.

James V. Neel, studying the most remote Indians of Central and South America, was struck by the lack of parental concern for the groveling play of crawling infants.[14] They frolicked happily in the debris of village and camp, dusty or mud-smeared, thrusting everything in their mouths, tasting their way, so to speak, into the environment. Neel knew that infants everywhere do the same if given the chance. But what to our sanitary-minded society looks like infantile perversity has the wisdom of enabling the child to begin building a repertoire of antibodies against the local antigens while still in part protected by the immunoglobulins of its mother's milk. To do so they had to "meet" the local antigens, and the result was that the children he examined had extremely high levels of antibodies against infection. In a somewhat different perspective the child was building into its physiology an immunological counterpart of the antigenic landscape, a mapping done by children the world over if given the opportunity.

The foregoing suggests that the habitat is not merely a container but a structured surround in which the developing individual makes tenacious affiliations, that something extremely important to the individual is going on between the complex structure of those particular places and the emerging, maturing self, a process of macro-micro correlation, mostly unconscious, essential to the growth of personal identity.

With this in mind, we can turn again to those large, excruciatingly detailed paintings made along the rivers and among the mountains of the Northeast during the second third of the nineteenth century. Up to the Revolution, the American knew himself in three contexts: as Christian, English colonial, and village community member. As this scaffolding was cut away by independence, secularization, and industrial-urbanization he suffered an acute attack of inchoateness from which he still has not recovered. The landscapes of those institutions had been the stable rural countryside tightly and hierarchically ordered around the church and town, making zones upon the land and interpenetrations with the wild that changed little between 1520 and 1820.[15] In an era of rapid change we may forget how constant was American life for three centuries, and we may overlook how traumatic the collapse of that old order was.

The main body of the Hudson River school is preceded by a literature of travel and description, but more significantly by an indigenous prose and poetry (even though imitative of European themes and styles). Pre-eminent are the works of Washington Irving, James Fenimore Cooper, and William Cullen Bryant. From time immemorial the myths of creation have been presented as epic tale, and perhaps could be comprehended only that way. The "legends" of Sleepy Hollow and the adventures of Natty Bumppo and Leatherstocking among the Indians were geographically explicit. It was possible, as John Trumbull did in 1810, to go to Norwich Falls in Connecticut, where Cooper had placed a climactic scene in *The Last of the Mohicans*, and do a portrait of the place, an analysis of its character the way one would analyze a person. Trumbull was the painter of heroes and battles, whose work hangs in the Capitol in Washington. He probably painted no more than three landscapes in a busy career. The fictional heroes and events, no less than the historical, gave place to elements of American identity—for we identified with the Irving and Cooper characters.

From the "actual" sites of such places, which much of the early painting scrutinized with an almost frenetic intensity, the painters moved out in search of correspondingly dramatic sites, appropriate to the imagination of episodes of pioneering life, or even storms or mountain geology. Thomas Cole was its most fervent spokesman. He painted and wrote long essays. To be lost in the wilderness he said, was the supreme experience. It was a way of seeking one's roots, a primitive regression. Neither the eroticism nor the adolescent emotionality of his work have gone unnoticed. This sentiment, for which later critics had only contempt, had its purposes, for it turned Americans back to maternal themes, to the land itself in a search for their beginnings, without which they would remain lost. Cynics after the Civil War saw those romantics merely as weak and undisciplined, as the Victorians considered women to be. To rhapsodize about trees and waterfalls seemed to them to have been sentimental and silly. The artists' own personalities were indeed rife with immaturity. But they were, in a sense, childish for us all.

Until then, the Europeans who settled America were on alien ground. With few exceptions their concrete connection to locality remained in Europe. The painters in America tried, in a few decades, to overhaul that whole troubled subjectivity, to imprint on our nervous systems a wild mountain spectacle as home, to do in the New World what had taken centuries to accomplish in Europe. Of course they could not possibly succeed, but in some ways no concept of place and landscape in America since then has been without something of their mark.

We are reminded with painful regularity of our continuing sense of dislocation, the neuroses of personal identity problems and the terror of loneliness in the crowd, of isolation both from society and from the rest of nature. These anxieties are linked to doubts about the purpose of life, even of order in the creation. Traditional psychology, scientific humanism, and even our religious preoccupation with the self have tried to explain these dilemmas of unconnectedness as arising within society and its works—in the family, the home, the job, or the church. But the failure of these explanations to either elucidate or remedy our chronic fragmentation raises doubts that our loneliness stems from inadequate social planning or ideology, or that we make or unmake ourselves apart from a nonhuman gestalt.

It is easy to blame rootlessness, mobility, and the fluidity of American life for our anguish, but all the hunting and gathering cultures that have ever been studied moved serenely through hundreds of miles without such troubles. Although they traveled through vast spaces, there is an organismic scale about their lives; W. H. Auden once observed that for us today the megaworld of the galaxy and the miniworld of the atom are real mainly in frightening ways. In "Ode to Terminus" he speaks of the Earth:

> where all visibles do have a definite
> outline they stick to and are undoubtedly
>   at rest or in motion, where lovers
>   recognize each other by their surface,
> where to all species except the talkative
> have been allotted the niche and diet that
>   becomes them. This, whatever micro-
>   biology may think, is the world we
> really live in and that saves our sanity.[16]

"Saving our sanity," in this mesocosm, might well require that we forget the heaven of the other-world religions along with the Adlerian psychology of simply willing your own world, as we have forgotten other tall stories.

Elide wants us to believe that places differ according to the amount of a universal holy oil we pour on them. He is in company with the cartographers, whose surveys of latitude, longitude, township, and range we have also accepted as the terms of location, of "defining space." But the world is not a billiard table until we finally turn it into one. It is unique everywhere in combining differently features that, in some unknown way, both reflect and create an inner geography by which we locate the self.

However exact the mathematical, political, or ecclesiastical subdivision of space may be, if it is imposed from a distant culture it cannot refer to place in the sense that is meant here, any more than maturity is achieved through ceremonies by which those institutions confer power on the individual, however much symbolic scenery they frame it in. British drama critic Tyrone Guthrie once wrote of Thornton Wilder's keen sense of place in his play *Our Town*:

> that such a close attachment to, and interpretation of a particular part of the earth is an absolute essential to any work of art which can ever be of deep or lasting significance. . . . It is one of the paradoxes of art that a work can only be universal if it is rooted in a part of its creator which is most privately and particularly himself. Such roots must sprout not only from the people but also the places which have meant most to him in his most impressionable years.[17]

Wilder not only creates place but an instrument of its recognition and affirmation. His own experience is one of discovery. He is like Carlos Castaneda, who tells us how difficult it is for one coming from a culture of the human domination of nature to discover, even in a room 12 by 8 feet, the spot on which he could sit without fatigue. His frustrating search under the tutelage of Don Juan took all night.[18] Again one thinks of the nineteenth-century painters, who roved on foot back and forth across the White and Green mountains, the Berkshires, the Taconics, the Hudson Highlands, and the Catskills, endlessly searching. The heterogeneity of the land is not made by humans—only discovered and enhanced or ignored and diminished by them.

An example of this reciprocity in which the given and the made play complementary roles is described in Jacquetta Hawkes's beautiful book *A Land*, an exercise in the gesalt-making powers where local rock is used for architecture, as it has been in many regions of England. The mind resonates between the stone as geology or terrain and its arrangement in man-made structures, between a unique regional architecture of the earth and the constraints and opportunities it offers in style and design. She is especially sensitive to the evolution of mind and consciousness, whose records occur as fossils in the same rocks that are used to build libraries, laboratories, and churches in which to contemplate the past.

Europe, as May Watts observed, had its cultural history written in the regional materials. Her guidebooks to the landscapes of Europe and America are achievements of a high order because she recognized that the value of place differences was not a matter of scenery but of the inter-action of people and their natural environment, a claim to which they cling despite the leveling bulldozer and premolded structures.[19] In America it is clearly less so than in Europe. True, there are the stone fences of Frost's New England and the grave slabs of Yankee marble, the limestone storefronts and banks of the Midwest, the sandstone campus at Boulder, and the adobes of New Mexico. Types of wood may also be identified with local architecture, though modern lumberyards have vir-tually lost their local connections. Wood does not last long enough to signify the earlier dependence on local trees in the way rock does. One is reminded of John Ruskin's refusal to come to America, saying that he could never visit a country that had no old castles. If you had no old castle you had no history, and if you had no history, there was no place in which the events that made you sanctified the ground.

But Ruskin was inordinately attached to the picturesque, to the neces-sity of ruins and the moral qualities of painting. Some dimensions of place do not depend on the interpenetration of geology and architec-ture. In her introduction to the poetry of Carl Sandburg, Rebecca West described the loquaciousness of Americans in public, their readiness to discuss their lives with total strangers, and the leisure they take in self-explanation:

> It occurs to one, as such experiences accumulate, that one has encountered in art, though not in life, people who talk and behave like this: in Russian novels. There one gets precisely the same universal addiction to self-analysis. And then it occurs to one also that this place is in certain respects very like Russia. Chicago, like Leningrad, like Moscow, is a high spot, to use its own idiom, on the monotony of great plains, a catchment area of vitality that rejoices extravagantly in its preservation because elsewhere in this region it might have trickled away from its source and been swallowed up in the vastness of the earth. All round Chicago lies the Middle Western plain. . . . The physical

resemblance between Russia and the Middle West is certainly close enough. And it may be that life which finds itself lost in the heart of a vast continent, whether that be Old or New, has a tendency to take the same forms. Life in another case, which flows in a number of channels and is divided into small nations, has an audience, who will give it a verdict on its performance, which is none the less useful if its inevitable function is to be disbelieved; and it has a basis for optimism about the universe, since it sees the neighboring nations surviving and flourishing in spite of what it is bound to consider their inferiorities. But a nation that is isolated in its vastness has no audience but itself, and it has no guarantee that continued existence is possible or worth while, save its own findings. Therefore, Russia and Middle West alike, it is committed to introspection, to a constant stocktaking of its own life and a constant search for the meaning of it.

In the Middle West more than anywhere else the introspective inhabitants have developed an idiom suited for describing the events of the inner life and entirely inadequate in dealing with the events of the outer life.[20]

In celebrating Chicago and the prairies Carl Sandburg threw the raw land right back in John Ruskin's face, for the prairie and Chicago were the least aesthetic places by the canons of painterly aesthetics, the concept of scenery.[21] The beauty of Missouri and Illinois will never come from matching them against the abstract standards of the picturesque, but from their affirmation by people in those places and by an integrity that is violated neither by alien aesthetics nor alien machinery. The commerce that tends to blenderize the world is not just physical, while its opponents are aesthetic. Modern tourism is not a defender of the world's difference against the utilitarian onslaught. Recreation, leisure, and art are the ministers of abstract scenery just as the centralized religions are of the abstract holy. Because of the abstract theory, a park might be established where nature happens to fit the standard. But the making of parks has been a license to surrender the rest.

Tourism and the park mentality, like that which pushed American Indians onto reservations, makes enclaves—not on the theory that quality is everywhere different, but on the theory that the alternative to dispersed points of sanctity is a continuous, weary uniformity. Sandburg was not signing parks and petrified village monuments like Lincoln's New Salem today. He belonged. Belonging, says Erik Erikson, is the pivot of life, the point at which selfhood becomes possible—not just belonging in general, but in particular. One belongs to a universe of order and purpose that must initially be realized as a particular society in a natural community of certain species in a terrain of unique geology. What Rebecca West sees as the empty plains of Illinois and Russia betrays her own bias, for they are empty only—as she noted—of close neighbors.

* * *

My theme can now be drawn together. It can be signified by the wandering of certain Australian aborigines. In going on the pilgrimage called walkabout, the Aborigine travels to a succession of named places, each familiar from childhood and each the place of some episode in the story of creation. The sacred qualities of each are heightened by symbolic art forms and religious relics. The journey is into the interior in every sense, as myth is the dramatic externalization of the events of an inner history. To the pilgrims these places are profoundly moving. The landscape is a kind of archive where the individual moves simultaneously through his personal and tribal past, renewing contact with crucial points, a journey into time and space refreshing the meaning of his own being.[22]

Terrain structure is the model for the patterns of cognition. As children we internalize its order as we practice "going" from thought to thought, and learn to recognize perceptions and ideas as details in the sweep of larger generalizations. We intuit these textures into a personal uniqueness. Mind has the pattern of place predicated upon it and we describe its excursions, like this essay, as a ramble between "points," the exploration of "fields," following "paths," and finding "boundaries," "wastelands," or "jungles," of the difficulty of seeing forests for the trees, of making mountains of molehills, of the dark and light sides, of going downhill or uphill.

Cognition, personality, creativity, maturity—all are in some way tied to particular gestalts of space, to locality partly given and partly found.

What does this say to us as Americans? From the standpoint of society as a whole, our disadvantages seem obvious and enormous. We have little cultural continuity with the land; history has few tangible relics. The vestiges of precolonial art and earthworks remain, but their meanings we do not know or feel. At the time of its settlement by Europeans the continent had vast diversity, as indicated by the diversity of Indian tribes. Almost everything we have done to it in the past two centuries has worked toward the destruction of these differences. We have idealized this uniformity in the image of the melting pot and the standard of living. The industrial complex levels mountains, drains swamps, opens forests, plants trees in grasslands, and domesticates and exterminates the wild. We have long been aware of this and of the rejoinder that it is a small price to pay for convenience, security, and comfort—that entertainment, travel, instant news, electrified homes, and an unlimited array of goods are made available in this way. Diversity, in fact, is suspect because it is divisive, or at best it is said to be just one more source of pleasure in a complex equation where the theme is "trade-offs." We are doubtful and ambivalent about diversity. Phyllis McGinley put it this way:

Since this ingenious earth began
  To shape itself from fire and rubble;
Since God invented man, and man
  At once fell to, inventing trouble,
One virtue, one subersive grace
Has chiefly vexed the human race.

One whimsical beatitude,
  Concocted for his gain and glory,
Has man most stoutly misconstrued
  Of all the primal category—
Counting no blessing, but a flaw
That difference is the mortal law.[23]

But the trouble is not that as a nation we lose the multifold character of a continent or lack architecture that affirms its diversity. We do not actually experience anything as a nation, but as individuals. To the corporation or bureaucracy the quality of place is merely an amenity, because our mythology of collective power confirms the transcendence of the individual.

The crucial point is that the child must have a residential opportunity to soak in a place, and that the adolescent and adult must be able to return to that place to ponder the visible substrate of his or her own personality. Place in human genesis has this episodic quality. Knowing who you are is a quest across the first forty years of life. Knowing who you are is impossible without knowing where you are. But it cannot be learned in a single stroke. This is what makes the commercial ravagement of the American countryside so tragic—not that it is changed and modernized, but that the tempo of alteration so outstrips the rhythms of individual human life.

In the 1830s it was said that hardly a Dutch house remained in Manhattan. Much of New England had been deforested by ax, fire, and merino sheep. A half-a-hundred painters went out in a furious attempt to find what might persist, attaching themselves to the most rugged terrain in their adolescent dismay and desire. In a hopeless frenzy of sheer will and stamina, they tried to establish grounds for self-discovery that would perhaps be stable, the way the European countryside was, across at least a generation of time.

Everywhere in America we continue to be engaged in that unspoken drama, to know the frustrations of being unable to grow up. Samuel Beckett, in his plays, has rightly set our quandary in an empty landscape, where we wait at crossroads marked only by signposts for something to happen, surrounded by a terrain that is both featureless and meaningless.[24] Signposts do not make a *whereness* nor beliefs of *whoness*.

If we were all as alike as eggs—and as eggs in our personal genesis we are unaware of our identities and potential relationships—it would not matter. But we hatch into a world where everything we do can help make or unmake the possibilities for our further growth. Intellectuals or eggheads like to think that we live in a world of ideas that we invent, as we create the domestic plants and animals. But in some part of our skulls there is wilderness. We call it the unconscious because we cannot cultivate it the way you would a field of grain or a field of thought. In it forces as enduring as climate and bedrock maintain our uniqueness in spite of the works of progress. It is the source of our private diversity. Together, our collective unconscious seems almost to exist apart from ourselves, like a great wild region where we can get in touch with the sources of life. It is a retreat where we wait out the movers and builders, who scramble continually to revamp our surroundings in search of a solution to a problem that is a result of their own activity.

NOTES

1. Vincent Scully, *The Earth, the Temple, and the Gods* (New Haven: Yale University Press, 1962).

2. Eric Isaac, "God's Acre," *Landscape* 14(2) (1964–65), and "Religion, Landscape and Space," *Landscape* 9(2) (1959–60).

3. Mircea Eliade, *The Sacred and the Profane* (New York: Harcourt, Brace, 1959).

4. Arthur Modell, *Object Love and Reality* (New York: I.U.P., 1968), and Bertram Lewin, *The Image and the Past* (New York: I.U.P., 1968).

5. Edith Cobb, *The Ecology of Imagination in Childhood* (New York: Columbia University Press, 1977).

6. A. R. Luria, *The Mind of a Mnemonist* (New York: Basic Books, 1968).

7. Edward Bennet, et al., *Journal of Neurobiology* 3:47 (1972).

8. Burton L. White, *Human Infants: Experience and Psychological Development* (New York: Prentice-Hall, 1971).

9. Grant Newton and Seymour Levine, eds., *Early Experience and Behavior* (Springfield: Ill., C. C. Thomas, 1968).

10. Michael Ellis, "Play: Theory and Research," in William Mitchell, ed., *Environmental Design: Research and Practice* (Los Angeles: University of California Press, 1972).

11. G. L. Johnson, *Proceedings of the Zoological Society of London* (1983), and *Philosophical Transactions of the Royal Society of London*, Series B, vol. 194 (1901).

12. Mayer Spivak, "Archetypal Place," *Forum* (October 1973).

13. Stephen Boyden, ed., *Impact of Civilization on the Biology of Man* (Toronto: University of Toronto Press, 1968).

14. James V. Neel, "Lessons from a Primitive People," *Science* 170:815 (1970).

15. J. B. Jackson, "The Westward Moving House," *Landscape* 2(3) (1973).

16. W. H. Auden, "Ode to Terminus," *The New York Review*, July 11, 1968.

17. Tyrone Guthrie, *The New York Times Magazine*, November 27, 1955, p. 27.

18. Carlos Castaneda, *The Teachings of Don Juan: A Yaqui Way of Knowledge* (Los Angeles: University of California Press, 1968), 18ff.

19. May Watts, *Reading the Landscape* (New York: Macmillan, 1957).

20. Carl Sandburg, *Selected Poems*, Rebecca West, ed., (New York: Harcourt, Brace, 1926).

21. Archibald MacLeish, "Where a Poet's From," *Saturday Review*, December 2, 1967.

22. Amos Rapoport, "Australian Aborigines and the Definition of Place," in Mitchell, ed., *Environmental Design*.

23. Phyllis McGinley, "In Praise of Diversity," *American Scholar* 23:306 (1954).

24. Robert M. Newcomb, "Beckett's Road," *Landscape* 13(1) (1963).

Human Nature

Ecology and Man—A Viewpoint

Ecology is sometimes characterized as the study of a natural "web of life." It would follow that man is somewhere in the web or that he in fact manipulates its strands, exemplifying what Thomas Huxley called "man's place in nature." But the image of a web is too meager and simple for the reality. A web is flat and finished and has the mortal frailty of the individual spider. Although elastic, it has insufficient depth. However solid to the touch of the spider, for us it fails to denote the *eikos*—the habitation—and to suggest the enduring integration of the primitive Greek domicile with its sacred hearth, bonding the earth to all aspects of society.

Ecology deals with organisms in an environment and with the processes that link organism and place. But ecology as such cannot be studied—only organisms, earth, air, and sea can be studied. It is not a discipline: there is no body of thought and technique that frames an ecology of man.[1] It must be therefore a scope or a way of seeing. Such a *perspective* on the human situation is very old and has been part of philosophy and art for thousands of years. It badly needs attention and revival.

Man is in the world, and his ecology is the nature of that *inness*. He is in the world as in a room, and in transience, as in the belly of a tiger or in love. What does he do there in nature? What does nature do there *in him?* What is the nature of the transaction? Biology tells us that the transaction is always circular, always a mutual feedback. Human ecology cannot be limited strictly to biological concepts, but it cannot ignore them. It cannot even transcend them. It emerges from biological reality and grows from the fact of interconnection as a general principle of life. It must take a long view of human life and nature as they form a mesh or pattern going beyond the conceptual bounds of other humane studies. As a natural history of what it means to be human, ecology might proceed the same way one would define a stomach, for example, by attention to its nervous and circulatory connections as well as its entrance, exit, and muscular walls.

Many educated people today believe that only what is unique to the individual is important or creative, and turn away from talk of populations and species as they would from talk of the masses. I once knew a director of a wealthy conservation foundation who had misgivings about the approach of ecology to urgent environmental problems in America because its concepts of communities and systems seemed to discount the individual. Communities to him suggested only followers, gray masses without the tradition of the individual. He looked instead— or in reaction—to the profit motive and capitalistic formulas, in terms of efficiency, investment, and production. It seemed to me that he had missed a singular opportunity. He had shied from the very aspect of the world now beginning to interest industry, business, and technology as the biological basis of their—and our—affluence, and which his foundation could have shown to be the ultimate basis of all economics.

Individual man *has* his particular integrity, to be sure. Oak trees, even mountains, have selves or integrities too (a poor word for my meaning, but it will have to do). To our knowledge, those other forms are not troubled by seeing themselves in more than one way, as man is. In one aspect the self is an arrangement of organs, feelings, and thoughts—a "me"— surrounded by a hard body boundary: skin, clothes, and insular habits. This idea needs no defense. It is conferred on us by the whole history of our civilization. Its virtue is verified by our affluence. The alternative is a self as a center of organization, constantly drawing on and influencing the surroundings, whose skin and behavior are soft zones contacting the world instead of excluding it. Both views are real and their reciprocity significant. We need them both to have a healthy social and human maturity.

The second view—that of relatedness of the self—has been given short shrift. Attitudes toward ourselves do not change easily. The conventional image of a man, like that of the heraldic lion, is iconographic; its outlines are stylized to fit the fixed curves of our vision. We are hidden from ourselves by habits of perception. Because we learn to talk at the same time we learn to think, our language, for example, encourages us to see ourselves—or a plant or animal—as an isolated sack, a thing, a contained self. Ecological thinking, on the other hand, requires a kind of vision across boundaries. The epidermis of the skin is ecologically like a pond surface or a forest soil, not a shell so much as a delicate interpenetration. It reveals the self ennobled and extended rather than threatened as part of the landscape and the ecosystem, because the beauty and complexity of nature are continuous with ourselves.

And so ecology as applied to man faces the task of renewing a balanced view where now there is man-centeredness, even a pathology of isolation and fear. It implies that we must find room in "our" world for all plants and animals, even for their otherness and their opposition. It further implies exploration and openness across an inner boundary—an ego boundary—and appreciative understanding of the animal in ourselves, which our heritage of Platonism, Christian morbidity, duality, and mechanism have long held repellent and degrading. The older countercurrents—relics of pagan myth, the universal application of Christian compassion, philosophical naturalism, nature romanticism, and pantheism—have been swept away, leaving only odd bits of wreckage. Now we find ourselves in a deteriorating environment, which breeds aggressiveness and hostility toward ourselves and our world.

How simple our relationship to nature would be if we only had to choose between protecting our natural home and destroying it. Most of our efforts to provide for the natural in our philosophy have failed—run aground on their own determination to work out a peace at arm's length. Our harsh reaction against the peaceable kingdom of sentimental romanticism was evoked partly by the tone of its dulcet façade but also by the disillusion to which it led. Natural dependence and contingency suggest togetherness and emotional surrender to mass behavior and other lowest common denominators. The environmentalists matching culture and geography provoke outrage for their oversimple theories of cause and effect, against the sciences that sponsor them and even against a natural world in which the theories may or may not be true. Our historical disappointment in the nature of nature has created a cold climate for ecologists, who assert once again that we are limited and obligated. Somehow they must manage in spite of the chill to reach the centers of humanism and technology, to convey there a sense of our place in a universal vascular system without depriving us of our self-esteem and confidence.

Their message is not, after all, all bad news. Our natural affiliations define and illumine freedom instead of denying it. They demonstrate it better than any dialectic. Being more enduring than we individuals, ecological patterns—spatial distributions, symbioses, the streams of energy and matter and communication—create among individuals the tensions and polarities so different from dichotomy and separateness. The responses, or what theologians call "the sensibilities," of creatures (including ourselves) to such arrangements grow in part from a healthy union of the two kinds of self already mentioned, one emphasizing integrity, the other relatedness. But it goes beyond that to something better known to twelfth-century Europeans or Paleolithic hunters than to ourselves. If nature is not a prison and earth a shoddy way-station, we must find the faith and force to affirm its metabolism as our own—or rather, our own as part of it. To do so means nothing less than a shift in our whole frame of reference and our attitude toward life itself, a wider perception of the landscape as a creative, harmonious being, where relationships of things are as real as the things. Without losing our sense of a great human destiny and without intellectual surrender, we must affirm that the world is a being, a part of our own body.[2]

Such a being may be called an ecosystem or simply a forest or landscape. Its members are engaged in a kind of choreography of materials and energy and information, the creation of order and organization. (Analogy to corporate organization here is misleading, for the distinction between social (one species) and ecological (many species) is fundamental.) The pond is an example. Its ecology includes all events: the conversion of sunlight to food and the food chains within and around it, man drinking, bathing, fishing, plowing the slopes of the watershed, drawing a picture of it, and formulating theories about the world based on what he sees in the pond. He and all the other organisms at and in the pond act upon one another, engage the earth and atmosphere, and are linked to other ponds by a network of connections like the threads of protoplasm connecting cells in living tissues.

The elegance of such systems and delicacy of equilibrium are the outcome of a long evolution of interdependence. Even society, mind, and culture are part of that evolution. There is an essential relationship between them and the natural habitat: that is, between the emergence of higher primates and flowering plants, pollinating insects, seeds, humus, and arboreal life. It is unlikely that a humanlike creature could arise by any other means than a long arboreal sojourn following and followed by a time of terrestriality. The fruit's complex construction and the mammalian brain are twin offspring of the maturing earth, impossible, even meaningless, without the deepening soil and the mutual development of savannas and their faunas in the last geological epoch. Internal complexity, as the mind of a primate, is an extension of natural complexity, measured by the variety of plants and animals and the variety of nerve cells—organic extensions of each other.

The exuberance of kinds as the setting in which a good mind could evolve (to deal with a complex world) was not only a past condition. Humans did not arrive in the world as though disembarking from a train in the city. We continue to arrive, somewhat like the birth of art, a train in Roger Fry's definition, passing through many stations, none of which is wholly left behind. This idea of natural complexity as a counterpart of human intricacy is central to an ecology of man. The creation of order, of which man is an example, is realized also in the number of species and habitats, an abundance of landscapes lush and poor. Even deserts and tundras increase the planetary opulence. Curiously, only man and possibly a few birds can appreciate this opulence, being the world's travelers. Reduction of this variegation would, by extension then, be an amputation of man. To convert all "wastes"—all deserts, estuaries, tundras, ice fields, marshes, steppes, and moors—into cultivated fields and cities would impoverish rather than enrich life aesthetically, as well as ecologically. By aesthetically, I do not mean that weasel term connoting the pleasure of baubles. We have diverted ourselves with litterbug campaigns and greenbelts in the name of aesthetics while the fabric of our very environment is unraveling. In the name of conservation, too, such things are done, so that conservation becomes ambiguous. Nature is a fundamental "resource" to be sustained for our own well-being. But it loses in the translation into usable energy and commodities. Ecology may testify as often against our uses of the world, even against conservation techniques of control and management for sustained yield, as it does for them. Although ecology may be treated as a science, its greater and overriding wisdom is universal.

That wisdom can be approached mathematically or chemically, or it can be danced or told as a myth. It has been embodied in widely scattered, economically different cultures. It is manifest, for example, among pre-Classical Greeks, in Navajo religion and social orientation, in Romantic poetry of the eighteenth and nineteenth centuries, in Chinese landscape painting of the eleventh century, in current Whiteheadian philosophy, in Zen Buddhism, in the worldview of the cult of the Cretan Great Mother, in the ceremonials of Bushman hunters, and in the medieval Christian metaphysics of light. What is common among all of them is a deep sense of engagement with the landscape, with profound connections to surroundings and to natural processes central to all life.

It is difficult in our language even to describe that sense. English becomes imprecise or mystical—and therefore suspicious—as it struggles with "process" thought. Its noun and verb organization shapes a divided world of static doers separate from the doing. It belongs to an idiom of social hierarchy in which all nature is made to mimic man. The living world is perceived in that idiom as an upright ladder, a "great chain of being," an image that seems at first ecological but is basically rigid, linear, condescending, lacking humility and love of otherness.

We are all familiar from childhood with its classifications of everything on a scale from the lowest to the highest: inanimate matter–vegetative life–lower animals–higher animals–humankind–angels–gods. It ranks animals themselves in categories of increasing good: the vicious and lowly parasites, pathogens, and predators–the filthy decay and scavenging organisms–indifferent wild or merely useless forms–good, tame creatures–and virtuous beasts domesticated for human service. It shadows the great man-centered political scheme upon the world, derived from the ordered ascendency from parishioners to clerics to bishops to cardinals to popes, or in a secular form from criminals to proletarians to aldermen to mayors to senators to presidents.

And so is nature pigeonholed. The sardonic phrase "the place of nature in man's world" offers, tongue-in-cheek, a clever footing for confronting a world made in man's image and conforming to words. It satirizes the prevailing philosophy of antinature and human omniscience. It is possible because of an attitude which—like ecology—has ancient roots, but whose modern form was shaped when Aquinas reconciled Aristotelian homocentrism with Judeo-Christian dogma. In a later setting of machine technology, puritanical capitalism, and an urban ethos it carves its own version of reality in the landscape, like a schoolboy initialing a tree. For such a philosophy nothing in nature has inherent merit. As one professor recently put it, "The only reason anything is done on this earth is for people. Did the rivers, winds, animals, rocks, or dust ever consider my wishes or needs? Surely, we do all our acts in an earthly environment, but I have never had a tree, valley, mountain, or flower thank me for preserving it."[3] This view carries great force, epitomized in history by Bacon, Descartes, Hegel, Hobbes, and Marx.

Some other post-Renaissance thinkers are wrongly accused of undermining our assurance of natural order. The theories of the heliocentric solar system, a biological evolution, and of the unconscious mind are held to have deprived the universe of the beneficence and purpose to which man was a special heir and to have evoked feelings of separation, of antipathy toward a meaningless existence in a neutral cosmos. Modern despair, the arts of anxiety, the politics of pathological individualism, and predatory socialism were not, however, the results of Copernicus, Darwin, and Freud. If man was not the center of the universe, was not created by a single stroke of Providence, and is not ruled solely by rational intelligence, it does not follow therefore that nature is defective where we thought it perfect. The astronomer, biologist, and psychiatrist each achieved for mankind corrections in sensibility. Each showed the interpenetration of human life and the universe to be richer and more mysterious than had been thought.

Darwin's theory of evolution has been crucial to ecology. Indeed, it might have helped rather than aggravated the growing sense of human alienation had its interpreters emphasized predation and competition less (and, for this reason, one is tempted to add, had Thomas Huxley, Herbert Spencer, Samuel Butler, and G. B. Shaw had less to say about it). Its bases of universal kinship and common bonds of function, experience, and value among organisms were obscured by preexisting ideas of animal depravity. Evolutionary theory was exploited to justify the worst in men and was misused in defense of social and economic injustice. Nor was it better used by humanitarians. They opposed the degradation of men in the service of industrial progress, the slaughter of American Indians, and child labor because each treated men "like animals." That is to say, men were not animals, and the temper of social reform was to find good only in attributes separating men from animals. Kindness both toward and among animals was still a rare idea in the nineteenth century, so that using men as animals could mean only cruelty.

Since Thomas Huxley's day the nonanimal forces have developed a more subtle dictum to the effect that "man may be an animal, but he is more than an animal, too!" The *more* is really what is important. This appealing aphorism is a kind of anesthetic. The truth is that we are ignorant of what it is like or what it means to be any other creature than we are. If we are unable to truly define the animal's experience of life or "being an animal," how can we isolate our animal part?

The rejection of animality is a rejection of nature as a whole. As a teacher, I see students develop in their humanities studies a proper distrust of science and technology. What concerns me is that the stigma spreads to the natural world itself. C. P. Snow's "Two Cultures," setting the sciences against the humanities, can be misunderstood as placing nature against art. The idea that the current destruction of people and environment is scientific and would be corrected by more communication with the arts neglects the hatred for this world carried by our whole culture. Yet science as it is now taught does not promote a respect for nature. Western civilization breeds no more ecology in Western science than in Western philosophy. Snow's two cultures cannot explain the antithesis that splits the world, nor is the division ideological, economic, or political in the strict sense. The antidote he proposes is roughly equivalent to a liberal education, the traditional prescription for making broad and well-rounded people. Unfortunately, there is little even in the liberal education of ecology-and-man. Nature is usually synonymous with either natural resources or scenery, the great stereotypes in the minds of middle-class, college-educated Americans.

One might suppose that the study of biology would mitigate the humanistic—largely literary—confusion between materialism and a concern for nature. But biology made the mistake at the end of the seventeenth century of adopting a *modus operandi* or lifestyle from physics, in which the question why was not to be asked, only the question how. Biology succumbed to its own image as an esoteric prologue to technics and encouraged the whole society to mistrust naturalists. When scholars realized what the sciences were about it is not surprising that they threw out the babies with the bathwater: the information content and naturalistic lore with the rest of it. This is the setting in which academia and intellectual America undertook the single-minded pursuit of human uniqueness and uncovered a great mass of pseudodistinctions such as language, tradition, culture, love, consciousness, history, and awe of the supernatural. Only we were found to be capable of escape from predictability, determinism, environmental control, instincts, and other mechanisms that "imprison" other life. Even biologists, such as Julian Huxley, announced that the purpose of the world was to produce man, whose social evolution excused him forever from biological evolution. Such a view incorporated three important presumptions: that nature is a power structure shaped after human political hierarchies; that man has a monopoly of immortal souls; and that omnipotence will come through technology. It seems to me that all of these foster a failure of responsible behavior in what Paul Sears calls "the living landscape" except within the limits of immediate self-interest.

What ecology must communicate to the humanities—indeed, as a humanity—is that such an image of the world and the society so conceived are incomplete. There is overwhelming evidence of likeness, from molecular to mental, between us and animals. But the dispersal of this information is not necessarily a solution. The Two Culture idea that the problem is an information bottleneck is only partly true; advances in biochemistry, genetics, ethology, paleoanthropology, comparative physiology, and psychobiology are not self-evidently unifying. They need a unifying principle not found in any of them, a wisdom in the sense that Walter B. Cannon used the word in *Wisdom of the Body*,[4] about the community of self-regulating systems within the organism. If the ecological extension of that perspective is correct, societies and ecosystems as well as cells have a physiology, and insight into it is built into organisms, including man. What was intuitively apparent last year—whether aesthetically or romantically—is a find of this year's inductive analysis. It seems apparent to me that there is an ecological instinct which probes deeper and more comprehensively than science and which anticipates every scientific confirmation of the natural history of man.

It is not surprising, therefore, to find substantial ecological insight in art. Of course there is nothing wrong with a poem or a dance that is ecologically neutral; its merit may have nothing to do with the transaction of man and nature. It is my impression, however, that students of the arts no longer feel that the subject of a work of art—what it "represents"—is without importance, as was said about forty years ago. But there are poems and dances as there are prayers and laws attending to ecology. Some are more than mere comments on it. Such creations become part of all life. Essays on nature are an element of a functional or feedback system influencing our reactions to our environment, messages projected by humans to themselves through some act of design, the manipulation of paints or written words. They are natural objects, like bird nests. The essay is as real a part of the community—in both the one-species sociological and many-species ecological senses—as are the songs of choirs or crickets. An essay is an Orphic sound, words that make knowing possible, for it was Orpheus as Adam who named and thus made intelligible all creatures.

What is the conflict of Two Cultures if it is not between science and art or between national ideologies? The distinction rather divides science and art within themselves. An example within science was the controversy over the atmospheric testing of nuclear bombs and the effect of radioactive fallout from the explosions. Opposing views were widely published and personified when Linus Pauling, a biochemist, and Edward Teller, a physicist, disagreed. Teller, one of the "fathers" of the bomb, pictured the fallout as a small factor in a worldwide struggle, the possible damage to life in tiny fractions of a percent, and even noted that evolutionary progress comes from mutations. Pauling, an expert on hereditary material, knowing that most mutations are detrimental, argued that a large absolute number of people might be injured, as well as other life in the world's biosphere.

The humanness of ecology is that the dilemma of our emerging world ecological crises (overpopulation, environmental pollution, etc.) is at least in part a matter of values and ideas. It does not divide us as much by our trades as by the complex of personality and experience shaping our feelings toward other people and the world at large. I have mentioned the disillusion generated by the collapse of unsound nature philosophies. The antinature position today is often associated with the focusing of general fears and hostilities on the natural world. It can be seen in the behavior of control-obsessed engineers, corporation people selling consumption itself, academic superhumanists and media professionals fixated on political and economic crisis, neurotics working out psychic problems in the realm of power over people or nature, artistic symbol-manipulators disgusted by anything organic. It includes many normal, earnest people who are unconsciously defending themselves or their families against a vaguely threatening universe. The dangerous eruption of humanity in a deteriorating environment does not show itself as such in the daily experience of most people, but is felt as general tension and anxiety. We feel the pressure of events not as direct causes but more like omens. A kind of madness arises from the prevailing nature-conquering, nature-hating, and self- and world-denial. Although in many ways most Americans live comfortable, satiated lives, there is a nameless frustration born of an increasing nullity. The aseptic home and society are progressively cut off from direct organic sources of health and increasingly isolated from the means of altering the course of events. Success, where its price is the misuse of landscapes, the deterioration of air and water, and the loss of wild things, becomes a pointless glut, experience one-sided, time on our hands an unlocalized ache.

The unrest can be exploited to perpetuate itself. One familiar prescription for our sick society and its loss of environmental equilibrium is an increase in the intangible Good Things: more Culture, more Security, and more Escape from pressures and tempo. The "search for identity" is not only a social but an ecological problem, having to do with a sense of place and time in the context of all life. The pain of that search can be cleverly manipulated to keep the status quo by urging that what we need is only improved forms and more energetic expressions of what now occupy us: engrossment with ideological struggle and military power, with productivity and consumption as public and private goals, with commerce and urban growth, with amusements, with fixation on one's navel, with those tokens of escape or success already belabored by so many idealists and social critics so ineffectually.

To come back to those Good Things: the need for culture, security, and escape are just near enough to the truth to take us in. But the real cultural deficiency is the absence of a true *cultus*, with its significant ceremony, relevant mythical cosmos, and artifacts. The real failure in security is the disappearance from our personal lives of the small human group as the functional unit of society and the web of other creatures, domestic and wild, which are part of our humanity. As for escape, the idea of simple remission and avoidance fails to provide for the value of solitude, to integrate leisure and natural encounter. Instead of these, what are foisted on the puzzled and troubled soul as Culture, Security, and Escape are more art museums, more psychiatry, and more automobiles.

The ideological status of ecology is that of a resistance movement. Its Rachel Carsons and Aldo Leopolds are subversive (as Sears recently called ecology itself).[5] They challenge the public or private right to pollute the environment, to systematically destroy predatory animals, to spread chemical pesticides indiscriminately, to meddle chemically with food and water, to appropriate without hindrance space and surface for technological and military ends; they oppose the uninhibited growth of human populations, some forms of "aid" to "underdeveloped" peoples, the needless addition of radioactivity to the landscape, the extinction of species of plants and animals, the domestication of all wild places, large-scale manipulation of the atmosphere or the sea, and most other purely engineering solutions to problems of intrusions into the organic world.

If naturalists seem always to be *against* something it is because they feel a responsibility to share their understanding, and their opposition constitutes a defense of the natural systems to which man is committed as an organic being. Sometimes naturalists propose projects too, but the

project approach is itself partly the fault, the need for projects a consequence of linear, compartmental thinking, of machinelike units to be
controlled and manipulated. If the ecological crisis were merely a matter
of alternative techniques, the issue would belong among the technicians
and developers (where most schools and departments of conservation
have put it).

Truly ecological thinking need not be incompatible with our place and
time. It does have an element of humility which is foreign to our thought,
which moves us to silent wonder and glad affirmation. But it offers an
essential factor, like a necessary vitamin, to all our engineering and social
planning, to our poetry and our understanding. There is only one
ecology, not a human ecology on one hand and another for the sub-
human. No one school or theory or project or agency controls it. For us it
means seeing the world mosaic from the human vantage without being
man-fanatic. We must use it to confront the great philosophical problems
of man—transience, meaning, and limitation—without fear. Affirmation
of its own organic essence will be the ultimate test of the human mind.

NOTES

1. There is a branch of sociology called human ecology, but it is mainly about
urban geography.

2. See Alan Watts, "The World Is Your Body," in *The Book on the Taboo against
Knowing Who You Are* (New York: Pantheon Books, 1966).

3. Clare A. Gunn in *Landscape Architecture*, July 1966, p. 260.

4. Walter B. Cannon, *Wisdom of the Body* (New York: W. W. Norton, 1932).

5. Paul B. Sears, "Ecology—A Subversive Subject," *BioScience* 14(7): 11 (July
1964).

Advice from the Pleistocene

Among the numerous explanations for the rebellions of college students is that the system reduces them to ciphers, professors do not know them as persons, and machines are depriving them of individuality.

This complaint has become a convention. Large student bodies and large classes, mechanization of record-keeping, professors' preoccupation with their research, mass feeding and housing seem to create a brave new world of personal nonentity.

As a convention it is readily exploited by viewers with alarm. Like most conventions it is a stereotype for a limited perspective. As a college teacher with experience in large and small institutions, I am convinced that numbers and machines are not the causes of student dissatisfaction, nor are they necessarily bad.

Most parents have no idea what they want a college to do for their children. Schools and children are so different from when they were young that they are wholly at sea when it comes to picking a school for Sarah or Johnny. They talk of academic excellence and well-rounded educations, but the truth is that they fear what Sarah will do when she is on her own.

Small colleges have exploited that anxiety for years. They deliberately created an image of the institutional family. The president is the daddy, and the dean of students is the mommy. Teachers are friendly uncles or big brothers.

During the past two decades, some colleges have perpetuated that image to their sorrow. Most of them are internally divided about the extent to which they should be substitute moms and dads.

But the damage has been done. There is now a large generation of adults who remember, with misplaced though profound sentiment, the intimate atmosphere of their college classes and kindly old Professor Warmheart, who took a personal interest in each of them.

Who are the students who want old Warmheart's attention today, and why? And why does he give it to them? My experience is that three out of ten students bring to him problems that are not even superficially academic. These problems old Warmheart has no business meddling in; they should be left to parents, professionals, or even friends of the students. Five out of ten are exploiting the opportunity for attention. They may keep the talk on Warmheart's special field of Mesozoic worms, but privately, and perhaps unconsciously, they are simply going through the motions of getting their money's worth. Two of the ten need his knowledge of worms, his scholarly advice, and the benefits of his experience.

And why is old Warmheart putting himself through all this? Surely he can distinguish the scholars from the neurotics and the ego-feeders after all these years?

He can. But he has his own problems and his own ego. If he has given up serious research or scholarship he has a big gap to fill. He is clever, articulate, and possibly self-deluded. The easiest, most deadly downhill road for the scholar is becoming a buddy to the students.

Fatuous and fatheaded Warmhearts, whose speciality is bull-slinging at an intimate level, are common though not limited to small colleges, which make a "thing" of small classes and individual attention.

As a teacher, I know that small classes are better than large classes—for certain subjects at certain times. But once a class enrollment exceeds seminar size—perhaps fifteen—it might as well be 1,500. As a student, I much prefer to hear a brilliant lecture in an auditorium than to participate in a windy discussion around a table. The place for small groups is in advanced courses and laboratories. My guess is that the proportion of these is as great as it ever was, in schools of all sizes.

What is missing in all the complaining about mass education is recognition of the virtues of anonymity. The Warmhearts, cozy social planning, and the familial aspects of academia add up merely to frippery. The opportunity to pursue the subject of Mesozoic worms to the limits of one's capacity, to stay up all night reading poetry, to experiment intellectually, socially, or sexually, have nothing to do with whether or not somebody is known to the institution and its functionaries as Sarah Tanner or as "74790."

What is so bad about being known in the registrar's office and the professor's book as "74790" instead of Sarah Tanner? We can readily imagine a Stone Age student at good old Neolithic U. complaining that she was no longer confronted as an individual because she was represented by symbols spelling her father's occupation: Tailor, Shepherd, or Tanner;

then, later, at Medieval Tech., we see the young Tanner girl bitter at the depersonalization forced on her by the addition to her name of one of a limited number of biblical names.

Names composed of letters now have ten centuries of literary-astrological superstition behind them. Number symbols for people have no such mystique going for them.

There are, to be sure, real problems in mass education, such as educationist puffery, pockets of decaying scholarship, administrative inertia, the entrenchment of uncritical thinking, and standardized tools. There are a few professors who do not want to talk to students—sometimes for justifiable reasons. But most professors in institutions of all kinds are deeply conscientious and sympathetic. Students must learn that their offices are not perpetual open houses. But I have never known a teacher who was vitally engaged with his subject to refuse to help a serious student—at a personal level.

Unless he has succumbed to fuddy-duddyism, the teacher gets very sharp at recognizing fakers and seekers for parental surrogates. He knows that the student out to master a subject is too busy to waste time at pseudodialogue, parading personal problems, or worrying about whether the college computer thinks of him in letters or numbers.

The Philosopher, the Naturalist, and the Agony of the Planet

The measure of surprise with which one reads Ortega from the perspective of anthropology or ecology can best be seen against the background of traditional European philosophy and historicism. These have tended, until recently, to reduce animals to machines or tissues and tribal peoples to savages. Derived largely from rationalist, analytical, and dualistic thought as exemplified in the work of René Descartes, the application of this inductive, objective approach to the natural and social sciences constitutes a cultural embedding only rarely challenged. This dominant Western view of man's relationship to nature created a hierarchy of values in which the nonhuman world and nonurban peoples are often seen as archaic components or factors at the most elemental level of existence or, worse, isolated from modern man qualitatively, so that little shared experience of life is possible.

It is not contended here that Ortega's work inevitably leads to a complete revision of the fragmenting and dualistic effects of a Western worldview; it is intended only to make a preliminary observation by a bystander on the possibility of an adjustment growing from his thought. Implicit in the study of ecosystems and man-nature integrations is the comprehensive view. Organismic and biological metaphors serve such conceptions. The naturalist is spontaneously attracted to see what sort of epistemological framework is intended, as in process philosophy, literary cosmology, gestalt psychology, evolutionary ethology, romantic organicism, or bioethological ecology.

It is often true in such integrative approaches that the organic model is not simply a happy choice of phrases but a literal part of the early experience of the scholar. Ortega's early training in biology and psychology is well known, but is remarkable in this respect, that the basic grounding in histology and physiology (and perhaps his reading of Goethe?), and the early ethology or "Umwelt" theory of Jakob Johann von Uexküll, together helped to prevent Ortega from getting stuck in the purely psychological aspects of perception and, ultimately, in idealistic and transcendental phenomenology. Equally important, the overview from the double standpoint of physiology and psychology lends itself to emphasis on the activity rather than the organs. In formulating reality not as a thing but as a perspective, Ortega opens the door to the notion of the world's concreteness as event rather than substance, a notion that may be the common thread of all holistic philosophy. To speak of bodies only as anatomy is to omit life. In the view of Charles Raven, that was exactly the mistake made by biology at the end of the sixteenth century, with the consequence that it vanished as a science independent of chemistry and physics until the publication of Charles Darwin's *Origin of Species by Means of Natural Selection.*[1]

An Ortega scholar has referred to the latter's declaration that "man must again search for the roots of his existence" as the central metaphor of his work.[2] Again and again the images of landscape as figures of the place and direction of thought, the underlying notion that each thing is a part of something larger, the quest for an authentic self as "being in the world" in generic terms, all testify to "giveness" as a beginning point. But "biologisms" surely permeate Ortega's thought far more profoundly than arbitrary literary analogies. From the naturalist's point of view this is a crucial point, at a time when "ecology" and its associated terminology have come to be widely used in the social sciences as synonyms for group processes, transformed into professional jargon by being stripped of their main thrust.

Ortega's notion of the concept as an organ of comprehension, its emphasis an act of "vital reason," is focused on a system of dyads in whose interrelation are events, past and present, constituting existence. Otherness is fundamental, as opposed to the mere idea of separateness. In insisting on the pullulating dialogue, even strife, between the organism and its surroundings, including other organisms, he skirts the trap set by logic for those who would find in it a mystical union, an "identity with nature" erroneously attributed in both romantic and cynical versions to the savage mind. In rejecting idealistic and eidetic phenomenology, Ortega avoids the multitude of reductionisms that crouch like lions to spring on holistic antelopes and reduce them either to chemical substances or to ideas.

As his theme develops, Ortega's vision of life as sport and perspective enlarges to include conflict in the process of self delineation. Here again, his corrective intuition heals an abused and misunderstood concept in descriptions of nature. He anticipates in broad terms what the ethological studies of Frank Darling, Charles Elton, Julian Huxley, Niko Tinbergen, and Konrad Lorenz have documented: that conflict and even violent death in nature are ritually treated within species, and in the service of finely tuned trophic systems between species.[3] The play of eater and eaten must be wrenched beyond recognition as explication or extension of human homicide and war. In natural systems the flow of nutrients and energy in parasitism, scavenging, and predation might almost be described as contractual, in which the psychology of individuals is characterized by accord. When Ortega speaks of the philosopher as hunter he makes the "totemic" conversion of trophic into social terms. That is, he employs an analogy of the kind which, according to Lévi-Strauss, was subverted by domesticated (or "caste") culture, impaired so that feeding systems in nature and historical human violence acquired literal continuity.[4] The popular sentiment against hunting is in part based on that confusion, which Ortega recognized as a monstrous misplaced tenderness, pride, and mannerism.[5]

In concert with Heidegger's concept of Being (*Dasein*) at the center of life is Ortega's assertion that being refers to the totality of every living thing's relation to all other things, and his exploration of the meaning of this assertion within the human frame. For example, to the question "What is the ideal theme of painting?" Ortega replies, "Man in Nature . . . man as an inhabitant of planet earth."[6] Such a statement is perhaps too easily misinterpreted in the cultural tradition of Western Europe and Anglo-America as meaning something like landscape painting by John Constable. It is essential that Ortega not be seen as evoking what might be called the puritan naturalism of Jonathan Edwards and John Muir. The redirection of mainstream philosophy from homocentrism, abstraction, and idealism has many possibilities other than the rubric of scenery, park, Peaceable Kingdom poetry or resource professionalism.

Indeed, much of the naturalism of that kind has been co-opted by the official Cartesian culture and given the status Art, Leisure, or Recreation to isolate and control it. That is, the West has preserved a schizoid division between the human and the natural to counteract the determinisms that it sees as axiomatic in evolutionary theory. That Ortega escapes this characteristic anxiety, which only renewed old dualisms, makes him of supreme interest at a time when the causes of planetwide destruction (of forests, soils, fresh water, native species, and watersheds) and pollution are being understood not only as short-sightedness or economic policy but also as ideology. No doubt, Ortega scholars are well aware of the details of the intellectual development that routed him around Hegelian dialectic and the Hobbesian biases of historical progress. As such, questions of this history of thought do not normally attract the naturalist. But the outfall from Ortega's own history could have a large impact on just those concerns of the naturalist and environmentalist who asks, "What are the educational and biographical clues which might provide insight into cultural changes of the kind that may be necessary for a new harmonious relationship between modern humans and their planet?"

Further, is it the case that once philosophy breaks free from the fear of reductionism in the perception of man as an organism in a world of organisms, it is forced to break new ground? If so, it would seem from the example of Ortega himself to involve the internalization of natural history and human prehistory in ways that do not conflict with what is perceived as uniquely human. It also has ontogenetic implications (and we note that Ortega was interested in the division of the individual life cycle into generic components). One of these is that categorical thought— that is, the capacity for cognitive grouping—emerges in concordance with speech in the child as is modeled on the concrete reality of species diversity, a diversity that provides, as Lévi-Strauss says, "the most available intuitive picture and direct manifestation of the ultimate discontinuity of reality," balanced by the cultural elaboration of tales and animals.[7] Childhood is a time of internalizing in active play and speech being an animal in order to, alternatively, become human.[8] The centrality of the verb *to be* is clear in Ortega's thought, just as it is a mode of cognition in childhood. Among adults in tribal societies this vital reasoning is perpetuated as an aural style. Such mode and style are not things but activities, soundings, movements like the flash of plumage in the forest. The categorical imagination is itself part of nature, which has its own hatching in the individual.

In this context, the Ortegan venture into the heart of the hunter may be seen as a reasonable pursuit of a philosophical anthropology that seeks the nature of the processes of becoming one's self. On one hand, *Meditations on Hunting* might seem to be a quaint, peripheral work, a kindness to an old friend who wanted a preface for his memoir. Its place in Ortega's total work cannot be appraised by a naturalist. But, on the other hand, it deals with a subject that is unusual among scholars. Perhaps its importance lies in part in its audacious theme. But then, how could a philosopher of nature write about pastoral scenery or the habits of squirrels who considers life to be "a grandiose and atrocious confluence, a formidable mystery" into which "hunting submerges man deliberately"?[9]

The quality of perception—the *attention* of the hunter as compared to that of other men—is one of the themes of that essay, a perspicacious and masterful insight.[10] Psychology, neurophysiology, and ethology have converged in recent years precisely in the analysis of attention.[11] This is the context in which the brains of higher animals, particularly man, can be understood as an evolutionary adaptation relating cognition and complex behavior through predator-prey relationships. As Ortega sees it, is on this ground that we can grasp the interconnections of the life of archaic humans, the lives of the diverse species and intelligent animals around them, and, ultimately, the sophisticated activity of the civilized philosopher.

Among his other insights into human ecology in *Meditations on Hunting* (indeed, anticipating such a field of study) are recognition of the importance of a conscious sense of generic human limitation: "If you believe that you can do whatever you like, even the supreme good, then you are irretrievably a villain;"[12] the biological complimentarity of predator and prey; the domesticated form of animals as pathological; the superficiality of some kinds of protectionism with their illusion of "saving" animals; the association of the hunt in human experience with homage, mystery, and divinity; acknowledgment of a widespread but nonsentimental yearning to be "in the country" and to recover "the pristine form of being a man"; the distinction between historical and generic human environments, and the effects of different ways of life on perceptual habits.

In his essay there are some ecological flaws, as in his description of the cause of the extirpation of the American buffalo. Probably there are also anthropological details that Ortega would report differently if he were writing today. In fairness, I should note that perhaps 90 percent of all the field studies of higher primates, wild predators, and hunting-gathering peoples have been done since 1950. In the essentials, however, and especially in anticipation of information not then scientifically documented, Ortega foresaw correctly. It is not the details but the focus on processes that relate, rather than on static things, in which he exemplifies what may in the long term be most helpful to the study of the relationship of man to nature.[13]

Victor Ouimette describes this attitude from *Meditaciones de Quijote:* "We reabsorb our circumstance by placing it within the context of a system of relationships. . . . Let us understand the hidden sense of all that surrounds us, everything that limits and thereby defines our life, without scorn for even the smallest realities."[14] The naturalist who spends his time looking at crickets or salamanders, things of no importance to the sweep of Western progress, may understandably feel a twinge of appreciation for such a remark. The naturalist's problem has been to formulate the significance of those minutiae in human terms; he desperately needs a philosophical rapprochement as much as traditional philosophy needs release from inertial, linear forms of thought toward a "sportive and festive sense of life."

A thread that links many who write in the literary tradition of the modern naturalist has often been a sort of humility, emphasizing relatedness to the natural order. In America, from Henry Thoreau to Loren Eiseley, such reflections have lacked the "depth" making their work meaningful to modern philosophers. The ideas prompted by such authors are, mostly, antithetic to the prevailing "sealed-off" temper of Christian, post-Renaissance thought. The naturalism of the nineteenth century notwithstanding, few writers with a strong, articulate sympathy for the importance of the Other have had available a theoretical foundation for raising their ideas to a plane whereby either formal theory or general premises of the culture could be affected. It seems that the present direction of the use of the earth in industrialized societies carries with it a philosophical vacuum. Consequently, there is emerging a resurgent, speculative address in ethics and aesthetics to this need, often characterized by internal tensions arising out of the conflict of the parent doctrines and the new feeling for wholeness. Like the work of Alfred North Whitehead, Ortega's thought has the advantage of embracing scientific information without accepting its reductionisms. In its biological embeddedness it avoids erecting hierarchic superstructures (unlike Whitehead), yet its emphasis on identity as the object of the task of being and its resonance with "circumstances" sustain awareness of a uniquely human situation without which metaphysics would remain largely physics.

There is now an opportunity for naturalists and philosophers (using these terms in the widest sense) to build upon Ortega's legacy, exploring new avenues to the human/environment which take into account the givenness of the human species. Historically, its advantage over its traditional antecedents are rootedness outside the Italian Renaissance and its scientific, humanistic European expressions. Insofar as Spain and Germany, exemplified by Ortega and Heidegger, insulated from the hubris and abstraction characterizing Latin culture, retain or recover an indigenous and more organic model of the complex of man and nature, they offer the seeds of perception and understanding that could contribute to a more harmonious interpenetration of the human and the Other.

NOTES

1. Charles E. Raven, *Natural Religion and Christian Theology* (Cambridge: Harvard University Press, 1953), 9.

2. Julian Marias, "The Metaphor" in *Ortega y Gasset: Circumstance and Vocation*, tr. Francis M. Lopez-Morillas (Norman: University of Oklahoma Press, 1970).

3. Julian Huxley, "Ritualization of Behaviour in Animals and Man," (symposium) *Philosophical Transactions of the Royal Society of London*, 251-B, 1956.

4. Lévi-Strauss, "Totem and Caste," in *The Savage Mind* (Chicago: University of Chicago Press, 1966).

5. Ortega y Gasset, "The Ethics of Hunting," in *Meditations on Hunting*, tr. Howard B. Wescott (New York: Scribner's 1972).

6. Ortega y Gasset quoted in Philip W. Silver, *Ortega as Phenomenologist* (New York: Columbia University Press, 1978), 29.

7. Lévi-Strauss, "Categories, Elements, Species, Numbers," in *The Savage Mind.*

8. Paul Shepard, "The Drama of the Animal," in *Thinking Animals* (New York: Viking, 1978).

9. Ortega y Gasset, "The Ethics of Hunting."

10. Ortega y Gasset, "The Hunter—The Alert Man," in *Meditations on Hunting.*

11. The role of attention as central to evolutionary theories of intelligence can be seen, for instance, in Harry J. Jerison, *Evolution of the Brain and Intelligence*, (New York: Academic Press, 1973), and in Monte Jay Meldman, *Diseases of Attention and Perception* (New York: Pergamon Press, 1970).

12. Ortega y Gasset, "The Ethics of Hunting".

13. Philip W. Silver, *Ortega as Phenomenologist*, 27, 102, 141. See also Victor Ouimette, *José Ortega y Gasset* (Boston: Twayne Publishers, 1982), 58.

14. Ouimette, *José Ortega y Gasset*, 57. Besides the sources above, my introduction to Ortega's thought has been greatly aided by the papers of Nelson R. Orringer and Sharon Hayes Nickel's "Ortega y Gassett," in "The Post-Marxian Critique of Industrial Society," Ph.D. dissertation, University of California, Los Angeles, 1974, and suggestions for reading by Professor Roberta Johnson of Scripps College

Hunting for a Better Ecology

Aldo Leopold's *Sand Country Almanac* was perhaps one of the most catalytic books in thinking on the environment in the first half of this century. Leopold died twenty-five years ago this spring, fighting a grass fire on his neighbor's lot, and accepted the first chair in game conservation in the United States, established by the University of Wisconsin just fifty years ago. *A Sand County Almanac* has become a small classic in this field. With Rachel Carson's *Silent Spring*, they are probably this century's two most incisive books on man and nature.

The *Almanac* went beyond questions of technique and how we are to improve the way we use our natural resources. Leopold was an ecologist, and yet the book transcended ecology, because the most noteworthy and perhaps most quoted chapters in *A Sand County Almanac* involved what Leopold termed "a land ethic" and "an ecological conscience." He saw these as more fundamental and necessary even than the management practices and exploitive manner by which Americans were skinning the continent. Leopold confronted us not only with a system that didn't work at the level of its instruments and its tools but a system that had not been working at the level of human conviction and human values and human spirit. In his emerging awareness Leopold has become for us a model of the change in attitudes that we now see as necessary, and yet a change that is difficult to describe and even more difficult to bring about in any kind of systematic way. Leopold, then, is the prototype, and a curious one, and we ask if this change is to happen in America how did it happen in Leopold himself? What happens to a person who brings this about, and is there anything there that in Leopold's own story might be helpful to us in coming to terms with so slippery a prescription as a land ethic?

Leopold was a trained forester and spent a large part of his early years with the U.S. Forest Service. He was an ecologist—that is, a biologist trained in the study of the interrelationship of organisms and their environment, and he was a hunter. And it's curious that if you look at the *Sand County Almanac* you find his pastime as a hunter becoming more and more important as a pivot on which his attitudes were changed.

Although the lessons he learned as a trained forester and ecologist were important, there were other foresters and other ecologists who failed then—and still fail—to recognize the difference between ethic and policy. Leopold's transformation comes from his hunting. He's a hunter who wrote in the preface to *A Sand County Almanac* that there are some people who can live without wild things and there are some who cannot, and that he was one of the latter. Perhaps this is an embarrassment to us because we're concerned with the conservation of life and we think of the hunter as a killer, not the gentle saviour, not the religious vegetarian or the religious metaphysician who deliberately goes about avoiding any killing of life. Here's a man whose own metamorphosis turned on the act of hunting. We're perhaps puzzled because we don't understand it and because our culture sees the hunter as that part of ourselves whom we'd just as soon forget—that past stage of crude savagery through which our ancestors had to pass to achieve civilization and rise to a higher humanity and understanding; that tradition in our society which thinks of the past in Hobbes's words as a "barbarianism," a matter of war of "all against all," a life which he said was "nasty, brutish, and short." It is regarded as a brief, unreflective interlude spent in constant fear and danger, in an environment that was always threatening either to destroy him directly by wild animals or an environment threatening to destroy by the lack of food or his exposure to the elements; a life of constant want, struggle, and deprivation, relieved finally only by modern life, where man was at the mercy of geography and weather as we are not, without control over his environment, and whose behavior was ruled, as it is in all cavemen, by his lowest emotions, his animal appetites and lusts. We admire perhaps his spirit to survive because, after all, it meant that we could come into existence, but we deplore his life as a life and his misfortune to have been born so long ago. So he is the animal in us, which we strive to overcome, that knows little of human love, and little of beauty as we understand it, or of rational thought—little, in other words, of civilized life.

The image of the caveman has for five thousand years been a spur to Christianity and to the goal of progress. Predation focuses on death—man as part of natural food chains—and it was necessary for him to kill other creatures to survive. Yet, it was the death of innocent things, and we are outrageously reminded by the savage of the past that we ourselves are stalked by death, and that we ourselves are cast in a natural mold from which we would escape into immortal life. Death is said to have come into the world with the Fall, and life to have existed very differently before Man sinned. One finds today many theologians who are much concerned about ecology and the environment; yet there is a limit to which they are ready to accept the ecological idea, and I think that that limit is very clearly seen in their resistance to an unavoidable order of existence whose pattern is based on food chains and the necessity of predator and prey, and of killing to eat. Many of my students and young friends likewise have reached this point in their reluctance to embrace ecology as a model of life. Ecology has been subversive not only as a challenge to a society that has infinite wants and is oriented to consumerism and pollution and compulsive manipulation but because its central reality is the food chain. Man's ecology is based on killing by hunting and gathering.

How true is the picture of the savage and barbarian that I've just described, how consistent with what we now understand about him from anthropologists who study the barbarian—from the examination of his bones and ancient campsites, as well as the hunter who still lives in some twenty-odd surviving relic cultures? For two million years, hunting-gathering was the shape and the frame of human life. That two million years provide us with a perspective on the ten thousand years since the beginning of agriculture and perhaps three or four thousand years since the beginning of urban civilization. Hunting, or cynegetic, man lived in small groups in a home range of several hundred square miles. His food consisted of about 20 percent meat. The hunter in our past—and in our present—is primarily a plant eater who eats meat as a relish, for whom hunting is more important in ceremonial and token ways than as nutrition. He lived and still lives in those parts of the land that are vegetatively parklike: glades, savannas, steppes, at the edge of the forest where forest and prairie meet and a mixture of plant and animal life is at its most rich and most diverse. Is he engaged in a daily and continuous struggle from dawn to dusk for his food? Studies of the Kalahari bushmen in Africa indicate that hunters and gatherers—the men doing the hunting, the women doing the gathering—work about two to three days a week, spending the rest of their time socializing, sleeping, dancing, visiting, being hosts, telling stories, playing with children, making music. The population size of these ancestral beings of our and of our cousins who still hunt and gather was quite small. Were they constantly afraid and in danger of being destroyed?

One of the most consistent observations of today's study of hunters is that they have no anxiety even about tomorrow. They lack our national epidemic of anxiety neurosis and widespread vascular disease. Both of these are caused by constant fear and tension and are not present in these people at any significant level. Are they indeed in danger all the time? There is a high mortality of infants but their life expectancy beyond the third year is similar to our own. They are not constantly being eaten by saber-toothed tigers or modern lions. There is no organized warfare.

What is the quality of cynegetic existence? Bushmen, aborigines, pygmies, and American Indians live or lived in a rich, stable, and diverse environment, using as many as fourscore kinds of plants for food and fifty or sixty kinds of animals. When one kind is unavailable it is easy enough to shift off to another. This constant environment is characterized by mature, natural communities—by which I mean a visible continuity, floods, storms, and weather are not so likely to devastate as they are more simplified environments of the kind associated with agriculture. Is he a club-carrying brute who is prepared to bash on the head the nearest person if he sees something he wants, and who has no understanding of the more artistic and sensitive side of human life? There is little evidence for it. Known cynegetic people spend a substantial part of each day engaged in talk, sleep, and music; some of the creations of Paleolithic hunters and gatherers are among the supreme achievements of mankind—I mean the drawings that still survive in the caves of France and Spain, Russia, Hungary, Italy, Czechoslovakia, Yugoslavia, and parts of North Africa.

What about his technology? Was he in fact dependent on a few blunt instruments, and is he still? The Aurignacian people, who lived twenty-five thousand years ago, had a tool kit of over two hundred kinds of instruments, many of which any individual man would have been capable of making. A number of these are so beautiful and exquisitely made that there are today only a handful of people in the civilized world who are able to make them—not only tools of stone, but of wood, bone, leather, tusk, and tooth. Is he in conflict with his neighbors and settling all of his arguments with his bows and arrows, his spears and his clubs? None of the living hunting and gathering people that we know carry on any kind of organized warfare or conscription of slavery, nor is there evidence that they did so in the past. We find warfare in an organized state among some peoples who still do some hunting and gathering, as in parts of central New Guinea, where there are no large mammals to hunt. Some of cynegetic man's encounters with neighboring peoples involve expiatory measures to remove possible friction by formality and ritual, but more often they are simply joyful meetings in which they see relatives and friends. They are bonded by the tradition of exogamy: the women marry outside of the immediate small band, so that they are related by blood, as it were, to the other people around them, who are not therefore alien. Are these people acquisitive? Do they gather and amass material goods? They can hardly do that, since they are largely mobile and they have only as much as they can carry. Generosity is the passway to prestige, and interestingly enough it's not simply that they *have* enough to engage in generosity, but that giving becomes intensified when things are scarce—when one kind of material, such as material for making into a tool, is in low supply. Are they diseased and malnourished? How can they be healthy without modern physicians and modern medicine and modern hospitals? One recent student of the Indians of Central America found them in superb health and considered that they were so because the children had been allowed to play in the mud and dirt and to get a large dose of bacteria while they were still quite young. They had extremely high antibody levels. And being creatures of sparse numbers and rare distribution, like wolves and tigers, contagious diseases are uncommon, as are epidemics. They are not in a good ecological position to pass on germs that sweep through populations of animals and humans in high density. Many of the illnesses of modern man, or any people in higher populations, come into existence or become important to our lives after our world population grew and the patterns of our distribution became such that illness could be easily passed on from one person to another. What of the savage's thought, sensibility, and attention: what quality of mind and consciousness had these barbarian, savage ancestors of ours? What was their normal attitude toward the surroundings they moved through, either as male hunters or female gatherers? According to the Spanish historian and philosopher Ortega y Gasset, the hunter's mind was finely tuned; Ortega, a contemporary of Leopold, never hunted but wrote a beautiful little essay on hunting that was translated by Howard Wescott. He says,

> "The hunter does not look tranquilly in a determined direction, sure beforehand that the game will pass in front of him. The hunter knows that he does not know what is going to happen,

and this is one of the greatest attractions in his occupations. Thus he needs to prepare an attention of a different and superior style, an attention which does not consist in riveting itself on the presumed, but consists precisely in not presuming anything and avoiding inattentiveness. It is a universal attention which does not ascribe itself on any point and tries to be in all points. We have a magnificent name for this, one that still conserves all the zest, vivacity, and imminence: alertness. The hunter is the alert man."

Ortega contrasts him to the farmer:

"The farmer attends only to what is good or bad for the growth of his cereals, or the maturation of his fruit. The rest remains outside his vision, and in consequence he remains outside the completeness that is the countryside. The tourist sees broadly the great spaces, but his gaze glides; it seizes nothing; it does not perceive the role of each ingredient in the dynamic architecture of the countryside. Only the hunter, upon imitating the perpetual alertness of the wild animal for whom everything is danger, sees everything and sees each thing functioning as facility or difficulty, as risk or protection."

Ortega knew nothing, or very little, of ecology. He may have not even heard the term at that point, because he doesn't use it in his essay. This being whom he describes is very unfamiliar to our culture. He forces our frame of past reference to change. My students, despite their openness to a new appraisal of the hunter, backslide once we get into a conversation about him. The old preconceptions and misconceptions always steal back into it in some way, because of our orientation to the idea of progress: the assumption that we have in fact improved things continuously and we need a kind of barbarian in the background shadows to convince us that we are better then we were. Cynegetic man is at a different past and different place than we know. Even if we accept him as perhaps a different being than we thought, of what possible interest can he be to the necessity of our own present and our own future? Even if we acknowledge that he is something beautiful, like an ancient tree or a past civilization, how are we to take this information into account in a world where obviously we cannot all be hunters and gatherers?

* * *

The answer is not easy, but there are some ways—in understanding human growth, for example. Our individual lives unfold in precise temporal ways, alike throughout our species, including hunting and gathering peoples. The life of each individual is framed in timed stages, and each step of this unfoldment has its environmental demands and transactions. Each of these came into existence as part of our evolution. We continually produce what produced us.

An example of this is from that part of the life cycle we call adolescence—let us say from the eleventh year to the sixteenth. This is the time when a young human being, in no matter what culture he lives, becomes interested in poetry and puns and in the new meanings of language. This is associated with the fact that he is going to receive the tales or myths of his culture, which tell in a metaphoric way, in a poetic form, those important truths about his own kind and his own purpose in the world. He is going to become disengaged from his family to some degree, and this disengagement is geared to a taciturnity and to the frictions between him and his parents which come about in almost every society. Disengaged from his parents and his siblings, he is going to become attached to a group and display an intensity of feeling for particular peers that he'll never again experience. Associated with this commitment is a kind of idealism in which the world is not nearly as good as it could be. Erik Erikson has called this the necessity of having something toward which one's fidelity can be directed. Hence he lives in a clique of blood brothers who are crucially important, for whom there exist certain mentors or guides or masters outside of his family. At fourteen a human becomes interested in ultimate questions in a way that he is not at ten or eleven. It is as though this unfolds following an inner blueprint in every individual, a stage whose wholeness must somehow now become complete and for which a feedback must be made, not only by society as a social organization, but as the translator of the ultimate Other, the whole of the cosmos. A confrontation must be arranged with his natural world or it will always seem alien, other, and potentially hostile to him. It's a time in which skill and at the same time experiences of solitude may be essential, and finally a time that culminates in some sort of initiation.

All the great modern religions recognize the importance of the initiation coming toward the end of adolescence, which solemnizes the transformation of this person into new status. That this is something invented by culture, I think, has been part of our ideological imagination. Rather, it seems to be built into the life cycle of the individual to require a return—a facilitating on the part of the natural world mediated by his society—in which the particular content can vary, but in which the basic necessity of receiving a poetic, mythic orientation about the world is given. It is part of his biology. Perhaps our whole life cycle eventually may be understood as characterized by these kinds of critical period needs, rather than obliging us to think of a human being as something that simply grows and expands until it's twenty-one and then starts down a staircase of decline and gradual failing of functions.

So we stand at the threshold of looking for new ways to direct and find a future that is more ecologically satisfactory and stable for us. There are clues, in our own background and in this kind of a past built into our own bodies, that must be taken into account, that will involve the redirecting of the human encounter with nature. We commonly hold that we want a society that operates economically and nearer to a steady state, a life that enables us to achieve personal fulfillment, enough small group contact to let us have a consistent body of close friends, the means for according a respect for all living things, a sense of the connectedness of life, of atunement and alertness. Such is a description of those primitive people. The wave of the future is the past. With those who say we cannot go back to the past, I would certainly agree. We cannot go back to the past, but the reason is not that we are caught on an arc that propels us through progress into a future. The reason is that we never left it. The genetic blueprint each of us carries in our cells is not some kind of gross outline or vegetative substructure but an exquisite set of options that may or may not be taken up by the social group and society and fulfilled, the kind of series of harmonic feedback, fields of activity, and action that demand specific timing and Otherness in the environment.

Instead we have created a degraded environment, not merely since industrialization, but beginning long before that. We have surrounded ourselves entirely with man-made things, including domestic animals, in which the diversity and richness of the natural and wild is diminished, deficient in what is essential for the life of the mind and in the wildness necessary for healthy development of attention. This pastoral world then is not a place of innocence and morality, as the eighteenth-century image of the Golden Age would have it. Agriculture was the origin of famine because of monocultures, where the life of the people depended on single crops, with their vulnerability to disease epidemics. The obsession about scarcity begins there, despite the fact that the world which replaced hunting and gathering was one that was occasionally more productive. A world emerged centered on the defense of place and territory and the aggressive expansion into territories in a way that was unknown to the people who preceded the farmers, a world that replaced a two-day week with a six-day week involving heavy seasonal drudgery and dulling, brutal routine from daylight to dark. Peasant existence is the price of civilization, where men struggle to retain their humanity and women become baby machines. This farming world includes those townspeople and urban parasites who grew up around the system of an agriculture.

How can we shift away from being world eaters and world destroyers, from making the earth's surface into a replica of our own desperate fears? What possible method or technology could one apply from any kind of wisdom about the past to our present situation, which seems to be dominated by an attitude developed during the past few thousand years? We already have the knowledge of the creation of synthetics—nonfood synthetics—that previously were produced on the land but would allow us to remove some 40 percent of the present cultivation of the earth and therefore to generate that much additional space, which could be included in wilderness. We have new microbes and know of yeasts and bacteria that can be fed upon human sewage, petroleum, and cellulose products in our environment, providing for us every needed nutrient—and which can be produced in much more restricted surroundings. This requires a sophisticated, elaborate technology new and different from anything man has done before, not going back simply to an economy of hunting and gathering as a way of life, or suggesting that we throw off our clothes and pass out spears at the door. By freeing the land, we could then move toward a recovery of the possibilities of being human.

To do this means abandoning of the hopeless destructive industrialization of the countryside and of agribusiness. By reorganizing human experience—separating what is man-made and cultural into highly polished and developed cities on one hand and wilderness on the other—we end the blurred areas of rural countryside and suburbs. In time we could recover a smaller population, but for the next century there is excellent evidence that the microbes we know of could support a world population of nine billion, three times the present world population, and still allow for the release of some 60 percent of the world's surface.

These techniques and this information are already available. It would take a bold and dramatic overturn of some of our thinking for us to apply it. Leopold himself took the bold step from ecology to philosophy. The example of his life says that philosophers all the way from Plato through Aquinas have used some word meaning "to hunt"—like *venator* or *thereutes*—to explain what they do. The philosopher, in the person of Ortega y Gasset, has also reminded us of this. He says: "The only man who truly thinks is the one who, when faced with the problem, instead of looking only straight ahead toward what habit, tradition, the commonplace, a mental inertial world would make one presume, keeps himself alert, ready to accept the fact that the solution might spring from the least foreseeable spot on the great rotundity of the horizon." The sturdy hunter in the field seems to us to be at the other extreme from the urbane scholar in his study, and yet there is a convergence of the thinking of these two men. But perhaps they have started to mend the split between the things of nature and the things of mind, not by unifying them, but by discerning relationships. Our task is reconnecting and reestablishing lines between them in rethinking the world of the future as something other than the trajectory of humanization of the whole planet. Leopold the hunter and Ortega the philosopher show that it can be done, and that what we can recover is not only a liveable world but the recovery of ourselves.

If You Care About Nature
You Can't Go On Hating
the Germans Like This

I learned about the Hun and the Bosche at my father's knee. The "enemy" for us small boys in our play with toy guns was just as often Germans as it was Indians. My uncle had a twenty-four-volume set of rotogravure picture books from World War I that provided abundant visual fantasy to feed the inner eye in our games of shoot and die. By the time I got to high school we were into World War II, and I served the last six months of the war in Europe with an armored division in France and Germany. My battalion "liberated" the Landsberg concentration camp, with its three living survivors, and of course, the horror of the Holocaust has not diminished in the decades since then. Even trivial experiences in later years—phalanxes of German tourists turning quiet Spanish cafés into loud cells of sentimental boorishness, or peaceful streets suffering the stench-in-the-ear of Volkswagens and Porsches—seemed to confirm the utter insensitivity of the "Kraut" wherever he appeared. In college I studied ecology (a word said to have been coined by the German zoologist Ernst Haeckel), and in class we made fun of the Prussian example of good/bad forestry: regimented plantings wherein the natural woodland community was so decimated that rodents swarmed in the absence of foxes and owls and destroyed the young trees. So, for a half-century the Germans seemed to me a dark and hopeless people, even though there is a family root on my mother's side of Schwartzes and Webers, ancestors of some distinction in Württemburg.

The other thread of this short discourse begins with that ecology course just mentioned. However mistaken the style of German forestry, we knew that ecology was to be applied to the care of the land. Particularly, I studied wildlife management, a program of applied science aligned in the land-grant universities with range management, soil management, forest

145

management, and so on. It seemed to us in those years, while Aldo Leopold was still living, that a great light was dawning. We understood that ecological relationships in nature were almost infinitely complicated, and yet the principle was simple—the whole of the living community had to be taken into account in the design of the various "managements." If that were done we would not go on making deserts, eroding the soil, and generally draining the earth of its resources. This was the idea or rubric under which all the earth skills would become a single enlightened resource management. In refined techniques for sustained production the habitat was to be considered. The managers were also to learn that their activities had side effects. We envisioned wildlife management as the prince of this new nobility of professions, since most of the game lived on land used or, hopefully, managed for something else.

Such were the conservation programs at mid-century. It was to be an educated exploitation in which the spoiler and waster would come to see that professionals so trained could enable them to get more from the earth in the long run. As a youth I had been interested in kinglets, fence lizards, and box turtles, and it occurred to me that these would not benefit much from management, however ecologically inspired. I had doubts about the scope of this approach to "wildlife" and wrote an article called "The Dove Is Doubtful Game," published by *Nature Magazine* (now defunct), which infuriated one of my management professors. To a colleague he used the word "traitor."

Feeling that something else was needed, I went off to an enlightened graduate school in the Ivy League. There we talked about cultural values and planetary matters like collapsing biomes, overpopulation, and biogeochemical cycles. Land-use techniques alone would not do. We needed to know about social processes, to see how the laws and taxes sometimes made land-skinners of the farmers or coal companies. We had to face up to the steely grip of habit and to the cults of use and waste, the status symbol of consumption and the psychology of growth and progress. Having given up chickadees for game animals and leaving them in turn for more distinctly human topics, I attempted to carve out a niche in this impossibly broad field by writing a thesis on the history of land aesthetics. What we needed, I supposed, was to balance out the practical use of nature against intangible values. Even the materialist West had a heritage of gardening, travel, and painting. The idea of the landscape fascinated me; it seemed to be the aesthetic equivalent of the ecosystem. If ecology was too esoteric for the public to understand (this was 1952), perhaps the notion of scenery was a built-in response to the same kind of thing.

It took me more than a decade to work my way through the landscape. I owe my liberation from it to the work of geographer David Lowenthal and social critic Marshall McLuhan. Their writing convinced me that the world-as-picture was, on one hand, geared to the superficiality of taste and, on the other, an outcome of a Renaissance mathematical perspective that tended to separate rather than join. Walter Ong's essay "The World as View and the World as Event" convinced me that this distinction between the visual and the tactile was more than ideological. The landscape was an inadequate nexus. It was only a twist in the idea of the co-option of the earth. Indeed, such ideas depended as much on unconscious perception as on intellectual or artistic formulations. I began to feel that something still more biogenic, yet common to humankind, which yet might take particular social or aesthetic expression, held the key to an adequate human ecology.

Over the next decade I read anthropology and child psychology. During that time a meeting of anthropologists took place in Chicago that resulted in the publication of *Man the Hunter*. I began to think that the appropriate model for human society in its earth habitat may have existed for several million years. If Claude Lévi-Strauss were to be believed, nothing had been gained by the onset of civilization except technical mastery, while what had been lost or distorted was a way of interpreting in which nature was an unlimited but essential poetic and intellectual instrument in the achievement of human self-consciousness, both in evolution and in every generation and individual human life. I knew such an idea would be ridiculed as a throwback to the discredited figure of the noble savage, but when it was considered in light of Erik Erikson's concept of individual development as an identity-shaping sequence I found it irresistible.

The essence of prehistoric human society could be better defined when seen against what replaced it: the peoples of plant and animal domestication, of agriculture, pastoralism, and the escalating of political entities, cities, and wars of the Near and Middle East, coupled with their catastrophic destruction of the land and their ascetic and abstract philosophies of transcendence in "world" religions. The subject was a bit large for a devotee of box turtles.

Still, there were intimations that might be worth passing along. Ideology and history do not seem very important in this connection in terms of their content—but ideologizing and historicizing are at its heart. They replaced the mythological foundation with its autochthonous roots and chthonic orientation. Therefore, I concluded, the focus should not be on management, good and bad ideas, nor even on the interpretive processes associated with them, but on something unidentified, on the mode or style of experiencing, something so fundamental that it had been poorly conceptualized. A destructive mode of being is somehow inflicted at an early age on every child in the Western world. One wants to call it Hebrew or Greek or Christian with Lynne White, but judging from the wreckage of the interiors of China and India, the Buddhists and Hindus offer no helpful alternative.

The needed corrective seems to me now to be northern. The poisoned ontology-ontogeny carried into Europe by Mediterranean cultures between the fall of Rome and the Reformation was diluted and resisted by the pagans and heathens who were, ostensibly, converted. The desert mind from the Mediterranean rim, a Platonic, prophetic, self-centering, dualistic, schizoid, eco-alienating way of being, could not have been less like the Celtic way that it eventually quashed and absorbed. As Japanese philosopher Watsuji Tetsuro has remarked, it is astonishing that a hemisphere of people in the north could believe that their whole existence hinged on things that happened to a small, distant, desert-fringe people two millennia ago. But the box-turtle reaction is to ask whether they have indeed come to believe it or whether ten millennia of tampering with the way we individually perceive ordinary experience and write history have not been managed so as to make it seem so.

The Protestants might be regarded as hard-liners who could see what the northern pagans were doing to Christianity, the compromises that the Church was making to gain a lease, the whole infiltration into the orthodox religion as described by Seznec in *The Survival of the Pagan Gods*. Like the Indians of North America centuries later, who went underground with their religion as the American government sought systematically to destroy their culture, the Britons, Finns, Hungarians, and Germans retained their "superstitions" in private and brought them masked to church. The prelates, in what they thought was a strategy of assimilation, kept the polytheistic holidays, but their success was their own perversion. It would be interesting to know today to what extent the good land-use—which makes much of Europe so beautiful—is due to customs that survived secretly, despite the otherworldly way that has dominated the North since the time of St. Augustine.

* * *

So now we come back to those hateful Germans with these thoughts: Techniques, however centrally organized or scientifically sophisticated, will not alone preserve the productivity and health of the planet. At a slightly deeper level, neither will alterations in the laws and governments, economic systems, political policies, or media, although they may yield remedial effects, save a species today, clean a river, reduce a smog, do some other temporary and piecemeal good. But they cannot even describe much less solve the problem. Going still further below these social realities, we may suppose that our behavior springs from philosophical systems or religious doctrines even in those of us who cannot articulate and seldom think about them. Such systems infuse the whole of a society with assumptions, interpretations, and implicit directions. Yet, for me, such a complex set of connections and pervasive outflow from reasoned theory do not explain adequately the way the youngest and uneducated members of society join so readily in its momentum and find alternatives so difficult to comprehend. A contemporary German philosopher who tells us that we must somehow find a still more basic mode below willed thought is Martin Heidegger.

So what is it about Heidegger that would make an Ozark ecologist forgive his German cousins? The professional philosophers cannot agree among themselves on what he says; therefore I can make whatever I can of that part of his work available in English and take some cues from George Steiner, the pungent stylist and author of a book on Heidegger. Primarily, it is that Heidegger makes constant connections to the biology of man in its broadest sense. Without building "up" from the natural sciences, he seems to move by a strange path and intuition toward certain psycho-ecological realities (in somewhat the same way that José Ortega y Gasset does); that is, by undercutting and rethinking many of the premises and givens of traditional Western thought.

He is obsessed with language and words—how they mean in a sense fundamental to experience and to our attention rather than as the external decoration of thought. His constant metaphors of movement through a terrain of the mind do not come across as arbitrary images of literary humanism but rather as a function or process of intelligence that evokes both a sense of origins in a deep past and the instrument by which each growing individual passes toward self-realization. It is as though the mental landscape is the lively shadow—no, not shadow, but aspect of a real place. Yet it is not a self- or even human-centered approach. Steiner remarks, "For Heidegger, on the contrary, the human person and self-consciousness are not the center, the assessors of existence. Man is only a privileged listener and respondent to existence. The vital relation to other-ness is not, as for Cartesian and positivist rationalism, one of 'grasping' and pragmatic use. It is a relation of audition. We are trying to 'listen to the voice of Being.' It is, or ought to be, a relation of extreme responsibility, custodianship, answerability to and for."

Heidegger insists on our awareness of the mystery of being and the numinous power of the words for being. His challenge to the thought of Aristotle, Plato, Bacon, Descartes, Hegel, Kant, and the other luminaries of the analytic and positivistic "willful sovereignty" of the ego is not even undertaken in the context of the transcendental Western metaphysics, and that, I suppose, is why it is so difficult. He seeks to oppose the "imperialist subjectivity" by seeking the defect in something prior to its logic.

"To rethink in this way," says Steiner, "a man must repudiate not only his metaphysical inheritance and the seductions of 'technicity,' but the ego-centric humanism of liberal enlightenment and, finally, logic itself. . . . True ontological thought, as Heidegger conceives it, is presubjective, prolog-ical, and above all, open to Being. It *lets Being be.* In this 'letting be' man does play a very important part, but it is only a part." Try that on your Buckmin-ster Fuller program.

In a world where the technician is no longer a craftsman, Heidegger is paying attention to "the existential fabric of everyday experience, the seamless texture of being which metaphysics has idealized or scorned. . . . Any artist, any craftsman, any sportsman wielding the instruments of his passion will know exactly what Heidegger means and how often the trained hand 'sees' quicker and more delicately than eye and brain."

Such an idea is for me like a crash of cymbals, triggering images of the marvelous Magdalenian tool kit, fifty or sixty or a hundred kinds of finely made instruments of bone, tooth, antler, stone, wood, thong, sinew, and skin that were replaced in Neolithic times by coarse implements in the deformed hands of men enslaved by shovels, harnesses, hoes, and plows. It helps me to understand what the poet Gary Snyder means by "work"—not a drudging, slogging subservience to a hated routine but an encounter with the world in a delicate, tactile way central to reflection and imagination.

Heidegger's feeling for the sanctity of the earth is not based on a penitential ecology, reparation, pseudotheology, or political radicalism, says Steiner; not an increase in knowledge or mastery but a "collaborative affinity with creation," which calls even for a renewal of ancient deities and the play of the agencies of vital order as they were seen in "pre-Socratic Greece."

* * *

For my part, the path Heidegger takes is not just the recovery of a rootedness, the personal and environmental integration among those who built the archaic Greek temples, which has so attracted generations of scholars, but something alluded to and never formulated in Lévi-Strauss's *Savage Mind*, which one may sense in the broken remains or visible tip of the submerged continent of American Indian reflective spirit, or some possibilities inherent in the cave art as described in André Leroi-Gourhan's *Treasures of Paleolithic Art*. It is not recoverable entirely by what is usually meant by philosophical speculation, philosophy being already prejudiced by its Classical inventors (though that may still be worthwhile to try). It may be latent in the way in which speech and the sublimation of sensory experience emerge in the life of the child, though it is not childlike. Somehow there is a wrong turning in what is done in childhood, an error repeated as each generation brings forth children—to paraphrase Winston Churchill, in the way we shape a world and it then shapes us.

References to a Heideggerian kind of earth, place, and instinct release in us a deep fear of Teutonic barbarism, the Hun horseman, memories of the slaughter of whole societies by invading pastoralists and warriors. I knew a Jewish academic traumatized by memories of the Holocaust who, forty years later, blocked the academic appointment of a gifted literary scholar for having studied ethology with Konrad Lorenz, whose name had been tainted by Nazism. The Celts and Gauls were for a millennium the complete examples of the Christian idea of the heathen.

How are we to understand this paradoxical German embodiment of the worst and the best? Attila, trees standing at attention, and the Prussian hobnailed mind apart, some of the worst is not northern at all, but rather an overlay brought north whose Latin tyrants defined barbarism according to their own best interests. King David, Constantine, and the Moorish caliphs showed few tender mercies in the massacre of whole populations in their day. Our society has revisioned savagery through the idealized, Mediterranean, desert-fringe lens. It is a bad lens that, in the interest of history, identifies the true barbarians as the others—a hypocritical lens. Indeed, in that northern ferocity there is a big contribution from hot minds and heads of adopted Near and Middle East ideologists—Zoroastrians, Muslims, Christians, and Jews—who themselves and their antecedents have been killing each other and other species with little enough reason for five thousand years.

Heidegger troubles our rationalizing mind in the same way that American Indians do, only he is more articulate about it, being his own primary source, not filtered through the ethnographic strainer. Only D. H. Lawrence, with his attention on the living continuum of wholes, ecologism, focus on place, repudication of "3,000 years of ideals, bodilessness, and tragedy," and overriding concern with the "fullness of being," compares to Heidegger in his effort to recover the healthy root of human circumstance.

* * *

As for me, I am going bird-listening. But I expect to hear from you young scholars what Heidegger is really saying. It wouldn't be at all surprising if he turned out to be the voice of a new and deeper environmentalism.

Virtually Hunting Reality
in the Forests of Simulacra

It would be difficult to argue with the assertions that our representations of the world are always "interpretations," that concepts shape our perceptions, that the human organism is its own shuttered window. Here I wish to explore the conclusion that reality is therefore invented by words, much the way Benjamin Whorf claimed seventy years ago that colors are a consequence of their differentiation by names. In its current form this idea further argues that such inventions are motivated by a struggle for power and that there is no Grand Truth beyond texts that are "allusions to the conceivable which cannot be presented."[1] I am also interested in the relationship of this inaccessibility of reality to "virtual reality."

According to the postmodern view, what most of us think of as simulations are only focused chatter about an unknowable external world. Even though they use living materials, the practitioners of "restoration ecology" may now find themselves, in this deconstructionist view, in the same boat with the museum curators making dioramas or habitat groups, who claim to be making artificial reproductions of the past or present world—and who are, therefore, merely engaged unconsciously in a sort of paranoid babble, lost in the vapors of their own imaginations without a compass or a satellite.

A 1973 essay in *Science* asks "What's Wrong with Plastic Trees?"[2] The essay hardly deals with the question it poses; primarily it is the last gasp of a spurious "argument" between "preservation" and "conservation." But the question is important. More recently the curator of the Devonian Botanical Garden, writing in *The Futurist*, points to fabricated lawns and polyester Christmas trees, which do not wilt in polluted environments or have to be watered, giving his blessing to the wonderful world of "artificial nature."[3] For years I have bedeviled my students with the issue of the validity of surrogates for living organisms and the enigma of the mind-set in which this kind of ambiguity arises, wherein the nature of authenticity and the authenticity of nature are riddled with qualification.[4] For a time I thought the issue could be clarified by examining the megaphysiology of exchange by natural trees with their environment, health-giving to the soil, air, and other organisms, as contrasted to poisoning the surroundings by the industrial pollution from making plastic trees. Now I see that such comparisons no longer matter, since what I said to them, my own psychobabble, is itself the subject.

Reality—You Can't Get There from Here

Looking behind the façades of grafted signification was the intent in 1949 of Marshall McLuhan's brilliant book *The Mechanical Bride*, which dissolved the rhetoric of magazine advertisements to expose the tacit messages concealed in the hypocrisy, presumptions, and deceit of the corporate purveyors of consumerism and our own lust to be seduced.[5] According to the current literary fashion, however, McLuhan himself can now be deconstructed and his own agenda shown to be just another level of presumption and the struggle for power.

If McLuhan still lived in the ocean of positivist naïveté, David Lowenthal, geographer, appeared on the new, dry shore of equivocal reality only thirteen years later. In an essay called "Is Wilderness Paradise Now?" he argues that the substantive reality of wilderness exists solely in the romantic ideas of it. Even its inhabitants are fictions, the noble savage having been the first victim of the "new criticism." Even the buffalo, Lowenthal says, "is only a congeries of feelings."[6] The buffalo is not some *thing* among other things—the cowbirds, Indians, pioneers, and you and me. It exists only as the feelings that arise from our respective descriptions.

Lowenthal's argument is reminiscent of the psychological conception of life as being locked inside a series of boxes and therefore precluding us from knowing anything but our own internal pulsations: coded patterns of electrochemical stimuli. At the level of cerebral axons, nuclei, and their glial cohorts the assertion that my world is more real than yours seems ridiculous. And suddenly we are back in Psychology 101, where we realize that all contact with "reality" is translated to the brain as neural drumbeats, nothing more, and in Psych 102, where the instructor titillates the class with the sensational observation that no old tree crashes down on an island if there is no ear to hear it. A thrill runs through the class, who will go on to Literature 101 to learn that the other impulses arise in words, a barrier behind which is a vast, unknowable enigma—or, perhaps, nothing at all.

Many of us—including me—may think of a photograph as the visual evidence of a past reality, so that certain events may then be recalled or better understood. But we are now confronted with the assertion that there is nothing "in" a picture but light and dark patches or bits of pigment—that the events to which we supposed such photographs refer are not themselves in the blobs. If something actually occurred it cannot be known. The result, in the case of photographs of starving people, is insensitivity to human suffering. This callous aesthetic is the object of Susan Sontag's anathema, identifying her as grounded—like McLuhan— rather than detached with the postmodern solipsists.[7]

Reflected light from actual events, focused through a lens onto a chemically sensitized plate, inscribing images that can be transferred by means of more light, lens, and photochemicals onto paper, does not, however, neutralize the "subject matter" even if it is a century old. The assay of such a photograph on formal grounds is a form of aesthetic distancing. This surreality of pictures, which denies the terms of their origin, reminds us that, according to the fashion, such a picture is a text, an impenetrable façade, whose truth is hidden.

Likewise, Hal Foster observes that recent abstract painting is only about abstract painting. Paintings are no more than the simulation of modes of abstraction, he says, made "as if to demonstrate that they are no longer critically reflexive or historically necessary forms with direct access to unconscious truths or a transcendental realm beyond the world —that they are simply styles among others."[8] As painting becomes a sign of painting, the simulacra become images without resemblances except to other images. Like the events that occasioned the photographs, the original configurations to which abstractions refer no longer have currency. Reality has dissolved in a connoisseurship of structural principles. A twentieth-century doubt has interposed itself between us and the world. "We have paid a high enough price for the nostalgia of the whole and the one," says Lyotard, and so allusion of any kind is suspect.

In my view this denial of a prior event is an example of what Alfred North Whitehead calls "misplaced concreteness." Paradoxically, the postmodern rejection of Enlightenment positivism has about it a grander sweep of presumption than the metaphysics of being and truth that it rejects. There is an armchair or coffeehouse smell about it. Lyotard and his fellows have about them no glimmer of earth, of leaves or soil. They seem to live entirely in a made rather than a grown world; to think that "making" language is analogous to making plastic trees, to be always on the edge of supposing that the words for things are more real than the things they stand for.[9] Reacting against the abuses of modernism, they assert that life consists of a struggle for verbal authority just as their predecessors in the eighteenth century knew that life was a social struggle for status or a technological war against nature. Misconstruing the dynamics of language, they are the final spokespeople of a world of forms as opposed to process, for whom existence is a mix of an infinite number of possible variations making up the linguistic elements of a "text."

Under all narrative we find merely more layers of intent until we realize with Derrida, Rorty, Lacan, Lyotard, and critics of visual arts that our role as human organisms is to replace the world with webs of words, sounds, and signs that refer only to other such constructions. Intellectuals seem caught up in the dizzy spectacle and brilliant subjectivity of a kind of deconstructionist fireworks in which origins and truth have become meaningless. Nothing can be traced further than the semiotic in which everything is trapped. The chain of relationships that orders a functional fish market, the cycle of the tree's growth, breath, decay, and death, the underlying physiological connections that link people in communities and organisms in ecosystems or in temporal continuity— all are subordinate to the arguments for or against their existence. The text—the only reality—is comparable only to other texts. Nothing is true, says Michel Foucault, except "regimes of truth and power." It is not that simulacra are good or bad replicas—indeed, they are not replicas at all; they are all there is! In a recent essay, Richard Lee characterizes this attitude as a "cool detachment and ironic distanciation," an eruption of cynicism caused by our daily bombardment of media fantasy, assaults on the "real," and consequent debasement of the currency of reality."[10] But I think it is not, as he concludes, simply a final relativism. There is no room even for relative truth in a nihilistic ecology.

The deconstructionist points with glee to the hidden motivations in these "falsifications" of a past and perhaps inadvertently opens the door to the reconfiguration of places as the setting of entertainment and consumption. This posture is not only a Sartrean game or artistic denial. It spreads throughout the ordinary world, where pictorial, electronic, and holographic creations; architectural façades for ethnic, economic, and historical systems; pets as signifiers of the animal kingdom; and arranged news are the floating reality that constitute our experience. We seem to be engaged in demonstrating the inaccessibility of reality.

In the past half-century we have invented alternative worlds that give physical expression to the denial of disaster. Following the lead and iconography of *National Geographic* magazine with its bluebird landscapes, and then the architecture of Disneyland happiness, a thousand Old Waterfronts, Frontier Towns, Victorian Streets, Nineteenth-Century Mining Communities, Ethnic Images, and Wildlife Parks have appeared. One now travels not only in space while sitting still but "back" to a time that never was. As fast as the relics of the past are demolished, whether old-growth forests or downtown Santa Fe, they are reincarnated in idealized form.

As the outer edges of cities expand, the centers are left in shambles, the habitation of the poor, or they are transported into corporate wastelands, which administer distant desolation as if by magic. To console the middle-class inhabitants and tourists, a spuriously appropriated history and cityscape replace the lost center with "Oldtown." In 1879, Thomas Sargeant Perry, having looked at Ludwig Friedlander's book on romanticism, wrote: "In the complexity of civilization we have grown accustomed to finding whatever we please in the landscape, and read in it what we have in our own hearts."[11] An example would be the "monumental architecture" of the rocky bluffs of the North Platte River as reported by the emigrants on the Oregon Trail in the 1840s.[12] Were Perry present today he could say that we make "whatever we please" out there and then announce it as "found."

Michael Sorkin speaks of this architectural "game of grafted signification . . . [and] urbanism inflected by appliqué" and the "caricaturing of places." Theme parks succeed the random decline of the city with their nongeography, their surveillance systems, the simulation of public space, the programmed uniformity sold as diversity. Condensed, they become the mall.[13] It is as though a junta of deconstructionist body-snatchers had invaded the skins of the planners, architects, and tour businessmen who are selling fantasy as history, creating a million Disney-lands and ever-bigger "events" for television along with electronic playsuits and simulated places in three-dimensional "virtual reality." Apart from the rarefied discourse and intimidating intellectualism of the French philosophes, their streetwise equivalents are already at work turning everyday life into a Universal Studios tour. It is not just that here and there in malls are cafés representing the different national cuisines but that the referent does not exist. Who cares about authenticity with respect to an imaginary origin?

The point at which the architectural fantasists and virtual realists intersect with intellectual postmoderns and deconstructionists is in the shared belief that a world beyond our control is so terrifying that we can—indeed, must—believe only in the landscapes of our imagination.

Amid the erosion of true relicts from our past, can we not turn to the museum? What are the custodians of the portable physical relicts doing? According to Kevin Walsh, museums now show that all trails lead to ourselves, create displays equating change with progress, and reprogram the past not so much as unlike ourselves but as trajectory toward the present.[14] The effect is like those representations of biological evolution with humankind at the top instead of the tip of one of many branches. In effect the museum dispenses with the past in the guise of its simulation, "sequestering the past from those to whom it belongs." Its contents are "no longer contingent upon our experiences in the world" but become a patchwork or bricolage "contributing to historical amnesia." Roots in this sense are not the sustaining and original structure but something adventitious, like banyan tree "suckers" dropped from the ends of its limbs. "Generations to come," Walsh predicts, "will inherit a heritage of heritage—an environment of past pluperfects which will ensure the death of the past."

Not long ago I was in the Zoological Museum at St. Petersburg, Russia, an institution that has not reorganized its exhibits according to the new fashion of diorama art—those "habitat groups": simulated swamps, seacoasts, prairies, or woodlands, each with its typical association of plants and animals against a background designed to give the illusion of space. There in Russia, among the great, old-fashioned glass cases with their stuffed animals in family groups, with no effort at naturalistic surroundings, I felt a rare pleasure. I realized that the individual animal's beauty and identity remain our principal source of satisfaction. When all members of the cat family, or the woodpeckers, are placed together, instead of feeling that I am being asked to pretend that I am looking through a window at a natural scene, I am free to compare closely related forms. Instead of an ersatz view I have the undiluted joy of those comparisons that constitute and rehabilitate the cognitive processes of identification. Nor does the museum display in St. Petersburg insidiously invade my thoughts as a replacement for vanishing woodlands and swamps by substituting an aesthetic image for noetics, the voyeur's superreal for the actuality.

New Dress for the Fear of Nature

Plastic trees? They are more than a practical simulation. They are the message that the trees they represent are themselves but surfaces. Their principal defect is that one can still recognize plastic, but it is only a matter of time and technology until they achieve virtual reality, indistinguishable from the older retinal and tactile senses. They are becoming *acceptable configurations*. No doubt we can invent electronic hats and suits into which we may put our heads or crawl, which will reduce the need even for an ersatz mock-up like the diorama. These gadgets trick our nervous systems somewhat in the way certain substances can fool our body chemistry—as, for example, the body failing to discriminate radioactive strontium 90 from calcium. (Strontium was part of the downwind fallout of atomic bomb testing, which entered the soil from the sky, the grass from the soil, the cows from the grass, the milk from the cows, children from the milk, and finally their growing bones, where it caused cancer.) As the art of simulacrum becomes more convincing, its fallout enters our bodies and heads with unknown consequences. As the postmodern high fashion of deconstruction declares that the text—or bits cobbled into a picture—is all there is, all identity and taxonomy cease to be keys to relations, to origins, or to essentials, all of which become mere phantoms. As Richard Lee says of the search for origins in anthropology, any serious quest for evolutionary antecedents, social, linguistic, or cultural ur-forms, has become simply an embarrassment.

But is this really new or is it a continuation of an old, antinatural position that David Ehrenfeld has called "the arrogance of humanism"?[15] Mainstream Western philosophy, together with the Renaissance liberation of Art as a separate domain and its neo-Classical thesis of human eminence, were like successive cultural wedges driven between humans and nature, hyperboles of separateness, autonomy, and control. It may be time, as the voices of deconstruction say, for much of this ideological accretion to be pulled down. But as down-pullers trapped in the ideology of Art as High Culture, of nothing beyond words, they can find nothing beneath "text." Life is indistinguishable from a video game, one of the alternatives to the physical wasteland that the Enlightenment produced around us. As the tourists flock to their pseudo–history villages and fantasylands, the cynics take refuge from overwhelming problems by announcing all lands to be illusory. Deconstructionist postmodernism rationalizes the final step away from connection: beyond relativism to denial. It seems more like the capstone to an old story than a revolutionary perspective.

Alternatively, the genuinely innovative direction of our time is not the final surrender to the anomie of meaninglessness or the escape to fantasylands but in the opposite direction—toward affirmation and continuity with something beyond representation. The new humanism is not really radical. As Charlene Spretnak says: "The ecologizing of consciousness is far more radical than ideologues and strategists of the existing political forms . . . seem to have realized."[16]

Life—or Its Absence—on the *Enterprise*

The question about plastic trees assumes that nature is mainly of interest as spectacle. The tree is no longer in process with the rest of its organic and inorganic surroundings: it is a form—like the images in old photographs. As the plastic trees are made to appear more like natural trees, they lose their value as a replacement and cause us to surrender perception of all plants to the abstract eye. Place and function are exhausted in their appearance. The philosophy of disengagement certifies whatever meanings we attach to these treelike forms—and to trees themselves. The vacuum of essential meaning implies that there really is no meaning. A highbrow wrecking crew confirms this from their own observations of reality—that is, of conflicting text.

There is a certain bizarre consequence of all this, which narrows our attention from a larger, interspecies whole to a kind of bedrock ethnos. For example, in each episode of the television series "Star Trek" we are given a rhetorical log date and place, but in fact the starship *Enterprise* is neither here nor there, now or then. Its stated mission to "contact other civilizations and peoples" drives it frantically nowhere at "warp speed" (speed that transcends our ordinary sense of transit), its "contacts" with other beings so abbreviated that the substance of each story rests finally on the play of interpersonal dynamics, like that which energizes most drama from Shakespeare to soap opera.

This dynamics is *essentially* primate—their own physicality is the only connecting thread to organic life left to the fictional crew and its vicarious companions, the audience. For all of its fancy hardware, software, and bombastic rhetoric, the story depends on projections of primatoid foundations, a shifting equilibrium like any healthy baboon social dynamic—the swirl of intimidation, greed, affection, dominance, status, and groupthink that make the social primates look like caricatures of an overwrought humanity.

The story implies that we may regress through an ecological floor—the lack of place and time and the nonhuman continuum—to farcical substitutes and a saving anthropoid grace. Should such a vehicle as the *Enterprise* ever come into existence, I doubt that their primatehood will save its occupants from the madness of deprivation—the absence of sky, earth, seasons, nonhuman life, and, finally, their own identities as individuals and species, all necessary to our life as organisms.

In the later series, "Star Trek: The Next Generation," the spacecraft has a special room for recreation, the "holo-deck" in which computers re-create the holographic ambience and inhabitants of any place and any time. The difference between this deck and H. G. Wells's time-travel machine is that the *Enterprise* crew is intellectually autistic, knowing that other places and times are their own inventions. The holo-decks that we now create everywhere, from Disneyworld to Old Town Main Streets, may relieve us briefly from the desperate situations we face on the flight deck as we struggle in "the material and psychic waste accumulating everywhere in the wake of what some of us still call 'progress.'" The raucous cries of post-Pleistocene apes on the *Enterprise* are outside the ken of the ship's cyborg second mate, just as the technophiles on earth will witness without comprehension the psychopathology of High Culture and literary dissociation, with its "giddy, regressive carnival of desire," rhetoric of excess, "orgies of subjectivity, randomness," and "willful playing of games."[17]

What, then, is the final reply to the subjective and aesthetic dandyism of our time? Given our immersion in text, who can claim to know reality?

As for "truth," "origins," or "essentials" beyond the "metanarratives," the naturalist has a peculiar advantage—by attending to species who have no words and no text other than context and yet among whom there is an unspoken consensus about the contingency of life and real substructures. A million species constantly make "assumptions" in their body language, indicating a common ground and the validity of their responses. A thousand million pairs of eyes, antennae, and other sense organs are fixed on something beyond themselves that sustains their being, in a relationship that works. To argue that because we interpose talk or pictures between us and this shared immanence, and that it therefore is meaningless, contradicts the testimony of life itself. The nonhuman realm, acting as if in common knowledge of a shared quiddity, of unlike but congruent representations, tests its reality billions of times every hour. It is the same world in which we ourselves live, experiencing it as process, structures, and meanings, interacting with the same events that the plants and other animals do.

NOTES

1. Jean-François Lyotard, *The Post-Modern Condition*, quoted in Jane Flax, *Thinking Fragments: Psychoanalysis, Feminism and Postmodernism in the Contemporary West* (Berkeley: University of California Press, 1990).

2. Martin H. Krieger, "What's Wrong with Plastic Trees?" *Science* 179:446–454 (1973).

3. Roger Vick, "Artificial Nature, the Synthetic Landscape of the Future," *Futurist* (July–August 1989).

4. See Umberto Eco, "The Original and the Copy," in Francesco J. Varella and Jean-Pierre Dupuy Crea, eds., *Understanding Origins* (Boston: Kluwer,1992).

5. Marshall McLuhan, *The Mechanical Bride: Folklore of Industrial Man* (New York: Vanguard Press, 1951).

6. David Lowenthal, "Is Wilderness Paradise Now?" *Columbia University Forum* (Spring 1964). See also my reply, "The Wilderness as Nature," *Atlantic Naturalist* (January–March 1965) .

7. Susan Sontag, *On Photography* (New York: Dell,1977).

8. Hal Foster, "Signs Taken for Wonders," *Art in America* (June 1986).

9. I used to think of this attitude as the "Marjorie Nicolson Syndrome." It was from her book *Mountain Gloom and Mountain Glory* that I first got the sense there were those who seemed to think the test of nature was whether it lived up to the literary descriptions of it. Clearly, Nicolson neither invented nor bears the full responsibility for this peculiar notion.

10. Richard B. Lee, "Art, Science, or Politics? The Crisis in Hunter-Gatherer Studies," *American Anthropologist* 94:31–54 (1993).

11. Thomas Sargeant Perry, "Mountains in Literature," *Atlantic Monthly* 44:302 (September 1879).

12. An account appears in Paul Shepard, "The American West," in *Man in the Landscape* (College Station: Texas A&M University Press, 1991).

13. Michael Sorkin, "See You in Disneyland," in Michael Sorkin, ed., *Variations on a Theme Park* (New York: Hill and Wang, 1992).

14. Kevin Walsh, *The Representation of the Past: Museums and Heritage in the Post-Modern World* (London: Roudedge, 1992).

15. David Ehrenfeld, *The Arrogance of Humanism* (New York: Oxford University Press, 1981).

16. Charlene Spretnak, *States of Grace* (San Francisco: Harper, 1991), 229.

17. Klaus Poenicke, "The Invisible Hand," in Gunter H. Lenz and Kurt L. Shell, eds., *Crisis of Modernity* (Boulder: Westview Press, 1986).

A Posthistoric Primitivism

The Problem of the Relevance of the Past

History as a Different Consciousness

H. J. Muller's classic *The Uses of the Past: Profiles of Former Societies* presented us with a paradox: "Our age is notorious for its want of piety or sense of the past. . . . Our age is nevertheless more historically minded than any previous age."[1]

Two decades later, with the publication of Herbert Schneidau's *Sacred Discontent*, the paradox vanished in a radical new insight.[2] For Schneidau, History was not simply a chronicle, nor even an "interpretation," but a new way of perceiving reality, one that set out to oppose and destroy the vision that preceded it. It does not refer to readers' understanding but to a cognitive style.

History, he said, is the view of the world from the outside. It was "invented" by early Hebrews who took their own alienation as the touchstone of humankind. Especially did they conceive themselves as outside the earth-centered belief systems of the great valley civilizations of their time. Central to those beliefs was cyclic return and its paradigmatic and exemplary stories linking past, present, and future with eternal structure. Schneidau calls this the "mythic" way of life. Alternatively, the view created by the Hebrews and later polished by the Greeks and Christians was that time may produce analogies but not a true embeddedness. All important events resulted from the thoughts and actions of a living, distant, unknowable God. There could never be a return. The only thing we could be sure of is that God would punish those deluded enough to believe in the powers of the mythic earth or who fell away from the worship of himself.

A perspective on Schneidau's concept of prehistory can be gained from recent studies of a style of consciousness among living, nonhistorical peoples. Dorothy Lee, describing the Trobriand Islanders, refers to the "nonlinear codification of reality"—space which is not defined by lines connecting points: a world without tenses or causality in language, where change is not a becoming but a new are-ness; a journey, not a passage through but a revised at-ness. Walter Ong calls it "an event world, signified by sound," a world composed of interiors rather than surfaces, where events are embedded instead of reading like the lines of a book. Of Eskimos, Bogert O'Brien says, "The Inuit does not depend on objects for orientation. One's position in space is fundamentally relational and based upon activity. The clues are not objects of analysis. . . . The relational manner of orienting is a profoundly different way of interpreting space. First, all of the environment is perceived subjectively as dynamic, experiencing processes. . . . Secondly, the hunter moves as a participant amidst other participants oriented by the action."[3]

For the Hebrews who invented History, the record of the linear sequence of ever new events would be the Old Testament. By the time we get to Herbert Muller that record has the density of civilized millennia and could be projected back upon the whole five thousand years of written words and such records as archaeology offers.

Muller's paradox, of our obsession with and obliviousness toward history, vanishes because we can begin to understand that the passion is an anxiety with our circumstances and our identity, which only grow thicker, like layers of limestone, as we burrow into that vast accumulation. The hidden truth of history is that the more we know, the stranger it all becomes. It is human to want to know ourselves from the past, but History's perspective narrows that identity to portraits, ideology, and abstractions to which nation-states committed human purpose. True ancestors are absent. Our search simply sharpens desire.

The meaning for our lives, of nature, of purposeful animals, of simple societies, of everything in this "past," is in doubt. We do not feel our ancestors looking over our shoulders or their lives pressing on our own. The past is the temporal form of a distant place. Our view is that you cannot be in two places or two times at once. I speak of this as a "view" in the sense of Ong's observation that the modern West is hypervisual, and my own conviction that what it considers a "view" is a perceptual habit. From this viewpoint we can see mere "oral tradition" as a nadir from which it was impossible to know that water in time's river runs its course but once and that you can no more recover the primordial sense of earth-linked at-homeness than a waterfall can run backward. And further, once we have shaken off that mythic immersion and put on the garment of dry History, we are unable to shed the detachment and skepticism that define the Western personality, embodied in the written "dialogues" that Robert Hutchins defined as the central feature of Western civilization.[4]

History not only envisioned, it created sense of the moment. Its content is sometimes delectable, sometimes horrible, but always irretrievable except as beads on the string from which we now dangle. It deals with an arc of time and of measured location; its creative principle being external rather than intrinsic to the world; deity as distant, unknowable, and arbitrary. Central to History is a subjectivity that also distances us from our ancestors.

The legacy of History with respect to primitive peoples is threefold: (1) primitive life is devoid of admirable qualities, (2) our circumstances render them inappropriate even if admirable, and (3) the matter is moot, as "You cannot go back."

"You can 't go back" shelters a number of corollaries. Most of these are physical rationalizations—too many people in the world, too much commitment to technology or its social and economic systems, ethical and moral ideas that make up civilized sensibilities, and the unwillingness of people to surrender to a less interesting, cruder, or more toilsome life, from which time and progress delivered us. This progress is the work of technology. When technology's "side effects" are bad, progress becomes simply "change," which is, by the same rote, "inevitable." Progress is a visible extension of the precognitive habit of History that influences concept and explanation by modulating understanding. It was not only the mathematicians, astronomers, and philosophers of the modern era who gave us the theoretical basis of progress.

All of these objections—and they seem insurmountable—seem to me to imply a deeper mind-set, which does not have to do with the content of history. It is more a reflex than a concept. We care little for its theories or inventions since the time of Francis Bacon or for the moods in Christendom that reversed the older view of things only getting worse.

Its true genesis lies in the work of Hebrew and Greek demythologizers. They created a reality focused outside the self, one that could be manipulated the way god-the-potter fingered the world. In rooting out the inner-directed, cyclic cosmos of gentiles and naïve barbarians, they destroyed the spiraled form of myth with its rituals of eternal return, its mimetic means of transmitting values and ideas, its role in providing exemplary models, its central metaphor of nature and culture, and most of all its function as a way of comprehending the past. It began the deconstruction of the empirical wisdom of earlier peoples and culminated in the monumental Western view of reality, whose central theme was the outwardness of nature.

Along with pictorial space and Euclidean time goes the phonetic alphabet as inadvertent "causes" of estrangement.[5] But these are not simply inventions of the postmedieval West. They are markers in the way the world is experienced. Their antecedents occur in the Bronze Age Mediterranean, where much of what we call "Western" has its roots.

Elsewhere I have tried to describe this history as a crazy idea, fostered not as a concept so much as the socially sanctioned mutilation of childhood, the training ground of perception, by the blocking of what Erik Erikson called "epigenesis."[6] But, whatever its dynamic, History alters not our interest in the past (witness Muller's observation that we moderns seem more interested than ever) but the work of attention itself, the deep current of precomprehension running silently beneath our spoken thoughts.

History and Ambiguity

If we attempt to recover the difficult and "distant" art of tool-flaking we may do so over the objections of modern rationality, which denies that the pterodactyl can fly because no one has seen it do so. That is, you cannot know the ancient technique. Not only does History define it as beyond access, but incomprehensible. History thinks its own process is an evolution separating us by our very nature from our past—medieval, Neanderthal, or primate.

Central to History is the notion of a fixed essence, an inner state that persists in spite of the contradictions of appearance, that our visible form not only fails to inform but can be made to deceive. Shifting appearance is dangerous; larval forms signify evil. The question of our primate or Neanderthal past cannot be addressed except as alternatives to our present identity. We are predisposed by the immense cultural momentum of History to dismiss such ambiguous assertions as one of a larger class of moot points in which categorical contradiction, the simultaneous reality of two opposing truths about ourselves, is denied.

Equally paradoxical is the matter of being in two times at once, even though our senses tell us that we are not today what we were yesterday.

This movement from one state or one thing to another is not so much a problem for human consciousness as for meaning. The liminal or boundary area of categories heightens cognitive intensity. In the historical world, such transformations have been handled by accepting reality as made up of fixed identities, oppositions, and beyond them, transcendent meaning, declaring one of the appearances to be illusory, or by seeing them as good and evil. In all cases except the last the surface or apparent contradiction is cast into doubt in favor of some deeper, hidden, more real reality. Mainly this problem has been met in the West by denying appearance—especially when it shifts or is a larval state—as the true identity and instead postulating essences and spirits within or seeking principles and abstractions as the enduring, unchanging reality, despite outward shape.

In non-Western, nonindustrial, and largely nonliterate (hence nonhistorical) societies, external form is dealt with quite differently. Edmund Carpenter cites our difficulty with the visual duck-rabbit pun as our loss of the "multiplicity of thought," a collapse of metaphor in a mind-set related to phonetic writing.[7] A. David Napier has traced the matter in elegant detail in connection with the ritual use of masks as the perpetual means of assenting to a universal principle of shape-shifting. Coupled with dance, this is humankind's central means of reconciliation with a world of changes.[8] The many shapes in such masked dances testify also to a world in which abstractions are given lively form. Ahistorical peoples usually live in worlds where power is plural, as in egalitarian small societies in which leadership is not monopolized but changing and dispersed. The concrete or given model for this discontinuity of emphatic and exemplary qualities is the range of natural species. To varying degrees the animals and plants are regarded as centers, metaphors, and mentors of the different traits, skills, and roles of people. In polytheistic worlds there is no omniscience and no single hierarchy, although there may be said to loom a single creative principle behind it all. Insofar as they model diversity and the polytheistic cosmos, the animals provide metaphors of forms and movements that can be brought ceremonially into human presence, as interlocutors of change. Their heads as masks, the animals in such rites become combinational figures created to give palpable expression to transitional states. The animal mask on the body of a person joins in thought what is otherwise separate, not only representing human change, but conceptualizing shared qualities, so that unity in difference and difference in unity can be conceived as an intrinsic truth. And some animals, by their form or habit, are boundary creatures who signify the passages of human life. Finally, in dance these bodies move to deep rhythms that bind the world and bring the humans into mimetic participation with other beings.

The sophisticated Greeks after the time of Pericles ridiculed these predications, and the Jews and Christians rejected them. The thinness of music and dance in temples, churches, and mosques indicates the minimalizing of what was and is basic to hundreds of different, indigenous religions marked by "mythic" imagination.

The nature of the primitive world is at the center of our dilemma about essence, appearance, and change. Since we are not now what we once were—we are not bacteria or quadruped mammals, or apish hominids, or primitive people living without domesticated plants and animals—the dichotomy is clear enough. We each know as adults that we are no longer a child, yet we are not so sure that our being doesn't still embrace that other self who we were. We are attached to that primitive way of understanding, of double being, in spite of our modern perspective. Depth psychology has led us to understand that this going back is going into ourselves, into what, from the civilized historical view, is a "heart of darkness." Clearly a threat of the loss of self-identity is implied, swallowed by a second nature which is hidden and unpredictable.

As born antihistorians, our secret desire is to explicate the inexplicable, to recover what is said to be denied. It is a yearning, a nostalgia in the bone, an intuition of the self as other selves, perhaps other animals, a shadow of something significant that haunts us, a need for exemplary events as they occur in myth rather than History. If not a necessity, it is a hunger that can be suppressed and distanced. The experience of that past is in terms of something still lived with, like fire, that still draws us. We cannot explain it, but it is there, made fragile in our psyche and hearts, drowned perhaps in our logic, but unquenchable.

It has been said that those who do not learn from history are doomed to repeat it, and yet by definition it cannot be repeated. Presumably such repetition means analogy. One does not really "go back," but merely discovers similar patterns. To ask the question in the perspective of prehistory: what are we to learn from history? The answer: history rejects the ambiguities of overlapping identity, space, and time, and creates its own dilemmas of discontent and alienation from Others, from nonhuman life, primitive ancestors, and tribal peoples. Failing to enact prehistory, we can live only in history, caught between captivity and escape, afflicted with Henry Thoreau's "life of quiet desperation," now called neurosis. Since history began, most people most of the time have lived under tyrants and demagogues (Mr. Progress, Mr. Collectivity, Mr. Centralized Power, Mr. Growthmania, and Mr. Technophilia). No empire lasts, and when states collapse, their subjects are enslaved by other states.

The crucial question of the modern world is "How are we to become native to this land?" It is a question that history cannot answer, for history is the denativizing process. In history "going native" is a madman's costume ball, a child's romp in the attic, a misanthrope's escape.

Unlike History, prehistory does not participate in the dichotomy that divides experience into inherited and acquired. Nor does it imply that our behavior is instinctive rather than learned. It refers us to mythos, the exemplifications of the past-in-the-present. Ancestors are the dreamtime ones, and their world is the ground of our being. They are with us still.

The real lesson of history is that it is no guide. By its own definition, History is a declaration of independence from the deep past and its peoples, living and dead, the natural state of being, which is outside its own domain. Indeed, History corrupts the imperatives of prehistory. What are the imperatives? What are we to learn from prehistory? Perhaps as Edith Cobb said of childhood, "The purpose is to discover a world the way the world was made."[9]

Savagery—Once More

After 2,500 years of yearning for lost garden paradises in Western mythology perhaps one of the most outrageous ideas of the twentieth-century was the advocacy of a hunting-gathering model of human life. Much of the world is still caught up in making a transition from an agrarian civilization. A writer for *Horizon* proclaims that "An epoch that started ten thousand years ago is ending. We are involved in a revolution of society that is as complete and as profound as the one that changed man from hunter and food gatherer to settled farmer."[10] He alerts us to the colossal struggle to go forward from the tottering institutions of agricultural life, and I am suggesting that we do move ahead to—of all things—hunting-gathering!

Among the problems that plague the "uses of the past," as H. J. Muller called it, the search for a lost paradise seems to resist the "facts" of history. One wonders whether it is even possible to write about the deep past without nostalgia, or without creating a world that never existed. Its images are a mix of dreams and visions, infantile mnemonics, ethnographic misinformation, and attempts to locate mythological events in geographical space and recorded history. History, indeed, is not exactly antimyth, dealing as it does with "origins" and recitations of the significant events of the past. But its "past" is radically different from the one shaping human evolution.

It was great fun working on a book on hunter-gatherer people in the early 1970s, because almost everything that the layperson generally thought to be true of them was wrong. In writing *The Tender Carnivore*[11] I tried to avoid the snare of idealism by disarming my critics in advance. I avoided the beatifying language of Noble Savagery, and I engaged Fons von Woerkom to draw chapter headings, as his art was anything but romantic. Even so, the incredulity with which it was greeted was puzzling. Looking back, I now see that the objection was not only that primitive life was inferior and irrelevant but, in the lens of historical memory, inaccessible.

For two centuries the ideology of inevitable change had set its values in contrast to fictional images of the lost innocence of the deprived and depraved savage. Forty years ago George Boas traced the history of that idea of the primitive over the past two thousand years, from early attempts to associate tribal peoples with biblical paradise to various views of perfection and the saga of evolving mankind.[12] For the Greeks anyone who lacked civil life in a polis and spoke incoherently (babbled) was a barbarian. Hostility to the idea that we have anything to learn from savages has as long a tradition as the dream itself. Skepticism about the full humanity of the Hyperboreans and Scythians among some Classical authors was opposed by the idealizing of the Celts, the Getae, and the Druids by Herodotus and Strabo.

The Christians got their ideas on prehistory from Plato's *Laws* via the Romans, which portrayed the pagans as childlike. Spanish endeavors to associate American Indians with European *Sylvestres homines*, the wild man, and the legacy of the Greek *barbori* have been reviewed by Anthony Pagdon. He makes some distinctions between Franciscan and Jesuit perception of the Indians—the Franciscan determined to destroy Indian culture to Christianize, and the Jesuits ignoring the "secular" side of the culture as irrelevant—an ironic twist on holism.[13] Oddly enough, it was the "unnaturalness" of the native peoples rather than their "naturalness" that justified decimation. Natural men, for example, did not eat each other.

In neo-Classical times Dr. Johnson observed that the hope of knowing anything about the people of the past was "idle conjecture." Horace Walpole derided antiquarians' fantasies. Locke and Hume gave us images of slavering brutes as an alternative to Rousseau's fictions of innocence and integrity. Admiral Cook's Polynesia would not look benign after the untamed sons of Adam did him in on the beach in Hawaii. The images were part of the heritage of the Roman idea of barbarians, the Christian notion of pagans, and eighteenth-century political philosophy of the benighted savage. Von Herder, Hegel, Compte, and Adelung all strove to disassociate humankind from the "laws of nature," to identify culture with History, to see conscious intellect identified with urban life, property, law, government, and "great art," as the final flowers in the human odyssey. The tradition continues. As M. Navarro said as late as 1924 of the South American Campa, "Degraded and ignorant beings, they lead a life exotic, purely animal, savage, in which are eclipsed the faint glimmerings of their reason, in which are drowned the weak pangs of their conscience, and all the instincts and lusts of animal existence alone float and are reflected."[14] Or closer to home is the testimony of historian Will Durant: "Through 97 percent of history, man lived by hunting and nomadic pasturage. During those 975,000 years his basic character was formed—to greedy acquisitiveness, violent pugnacity, and lawless sexuality."[15] Quite apart from anthropology this conglomerate idea of the primitive remains the central dogma of civilization held by modern humanists.

By the end of the nineteenth century there emerged in the United States a substantial body of admiration for Indian ways. I remember as a boy in the 1930s meeting Ernest Thompson Seton in Santa Fe. He ran a summer camp at his ranch to which boys came to be tutored by local Navajos, bunk in tepees, and live out the handcraft and nature study ventures of *Two Little Savages*.[16] The image of the American Indians in this dialectic has been reviewed by Calvin Martin, who observes that by the late 1960s the image of the "ecological Indian" was being articulated by Indians themselves, notably Scott Momaday and Vine Deloria. Arrayed against them in postures of "iconoclastic scorn" are experts who pursued an old line in anthropological guise—debunkers of the image of the Noble Savage, which they said merely masked a knave who was not nature's friend but who typically overkilled the game at every opportunity.[17]

Oddly enough, science did not rapidly resolve what seemed to be a question of facts. Geology after Lyell, evolution after Darwin, and archaeological time after Libby's atomic dating complicated but did not settle much. With a slight twist evolution could be the handmaiden of Progress. "It began to look," says Glyn Daniel, "as if prehistoric archaeology was confirming the philosophical and sociological speculations of the mid-nineteenth-century scholars."[18] Anthropology idealized value-free science and cultural relativism, thwarting European chauvinism but throwing out the baby with the bath water.

I was, of course, not the first to try to formulate the meaning of hunting-gathering for our own time. But not all efforts to clarify the description of hunters were applied to ourselves. Knowledgeable writers tiptoed among the ferocious critics, pretending that hunting signified only a remote past, as in Robert Ardrey's *Hunting Hypothesis*[19] or John Pfeiffer's *The Emergence of Man*.[20] Nigel Calder's *Eden Was No Garden*[21] and Gordon Rattray Taylor's *Rethink*[22] stirred the pot, but could hardly be said to have influenced, say, the civilized dogma of the modern university. Scholarly silence greeted the English translation of Ortega y Gasset's *Meditations on Hunting*,[23] as though an imposter had inserted an aberration in his works.

The message is clear: Advocacy of a way of life that is both repulsive and no longer within reach seems futile. Time is an unreturning arrow. The hunting idea is a barbaric atavism, unwelcome at a time when aggression and violence seem epidemic. The idea is obviously economically impractical for billions of people and incongruent with the growing concern for the rights of animals. Animal protectionists and many feminists seem generally to feel that hunting is simply a final grab at symbolic virility by insensitive, city-bred male chauvinists, or one more convulsion of a tattered and mis-placed nostalgia. Less and less, however, is hunting condemned as the brutal expression of tribal subhumans, for that would conflict with modern ethnic liberation.

* * *

The idea of inherent "nobility" of the individual savage was laughed out of school a century ago, properly so. Hunter-gatherers are not always pacific (though they do not keep standing armies or make organized war) nor innocent of ordinary human vices and violence. There is small-scale cru-elty, infanticide, and inability or unwillingness to end intratribal scuffling or intertribal vengeance. From the time of Vasco da Gama, Westerners have been fascinated by indigenous punishment for crimes and by canni-balism (although cannibalism is primarily a trait of agri-cultures). Hunter-gatherers may not always live in perfect harmony with nature or each other, being subject to human shortcomings. Nor are they always happy, content, well fed, free of disease, or profoundly philosophical. Like people everywhere they are, in some sense, incompetent. In *Little Big Man*, Indian actor Dan George did an unforgettable satire on the wise old chief who, delivering his rhetoric of joining the Great Spirit, lies down on the mountain to die and gets only rain in the face for his trouble. Given a century of this kind of scientific dis-illusioning, what is left?

It has been uphill and downhill for the anthropologists all along. The nineteenth-century "humanist anthropologists" like Edward Tylor and Malinowski dismissed native religious rites as logical error, although they allowed that ritual may work symbolically. As to the veracity of their religion, an "embarrassed silence" has marked anthropology ever since, say, Bourdillon and Fortes.[24]

Against these relativists there has also been an eccentric group of anthropologists who were not neutral about the tribal cultures. A. O. Hallowell, W. E. H. Stanner, Carleton Coon, and Julian Steward walked a narrow line between science and advocacy. Claude Lévi-Strauss rescued the savage mind. Coon's courage was exemplary. He scorned the "academic debunkers and soft peddlers," including those who spoke of "the brotherhood of man" as contradicting the reality of race.[25] Stanner was perhaps the most eloquent, describing aboriginal thought as a "metaphysical gift," its idea of the world as an object of contemplation, its lack of omniscient, omnipotent, adjudicating gods—a world without inverted pride, quarrel with life, moral dualism, rewards of heaven and hell, prophets, saints, grace, or redemption. All this among Blackfellows, whose "great achievement in social structure" he said was equal in complexity to parliamentary government, a wonderful metaphysic of assent and abidingness, "hopelessly out of place in a world in which the Renaissance has triumphed only to be perverted and in which the products of secular humanism, rationalism, and science challenge their own hopes."[26] If any modern intellectuals read him they must have thought he had "gone native" and left his critical intelligence in the outback.

After twenty centuries of ideological controversy it may be impossible to enter the dialogue without trailing some of its biases and illusions. But there is perspective from different quarters—from the study of higher primates, hominid paleontology, Paleolithic archaeology, ethology, ecology, field studies of living hunter-gatherers, and direct testimony from living hunter-gatherers.

A turning point was a Wenner-Gren symposium in Chicago and its publication as *Man the Hunter* in 1968.[27] The essays reported scientific evidence that the caveman as well as the noble savage was so much urban moonshine. It was a meeting of field-workers who had studied living tribal peoples in many parts of the world, coming together and finding common threads that linked diverse hunter-gatherer cultures to one another and to Paleolithic archaeology. This shift toward species-specific thinking benefited from "the new systematics," an evolutionary perspective based on genetics and natural selection articulated by G. G. Simpson, Ernst Mayr, Julian Huxley, and others. *The Social Life of Early Man*[28] was indicative of the new level of continuity among primitive societies, afterward given cross-cultural generalizations in George Murdock's ethnographic atlas.[29]

Although a few bold voices had been heard among them, such as Marshall Sahlins in *Stone Age Economics*,[30] their own evidence did not make anthropologists into advocates of a new primitivism. Their restraint was no doubt the result of a hard-won professional posture, the twentieth-century effort to overcome two centuries of ethnocentrism. But it was also the outwash of three generations of cultural relativism by mainstream social science, pioneered by Boas and Kroeber,[31] recently voiced with imperious assurance by Clifford Geertz, who wrote that "there are no generalizations that can be made about man as man, save that he is a most various animal."[32] Catch them saying that any culture is better than another!

In any case, such a judgment would be irrelevant, since even present-day hunter-gatherers are, by its historical logic, part of an irrecoverable past. Melvin Konner, a Harvard-bred anthropologist who spent years studying the !Kung San of the Kalahari Desert of Africa, wrote a fascinating account of his study showing the marvelous superiority of their lives to their counterparts in Cleveland or Los Angeles, and then pulled the covers over his head by saying, "But here is the bad news. You can't go back."[33] One can only be grateful for Loren Eiseley[34] and Laurens van der Post[35] in their admiration of the same Kalahari bushmen. Perhaps they anticipated what Roger Keesing calls the "new ethnography," which seeks "universal cultural design" based on psychological approaches. "If a cognitive anthropology is to be productive, we will need to seek underlying processes and rules," he says, observing that the old ethnoscience has been undermined by transformational linguistics and its sense of "universal grammatical design." He concludes that "the assumption of radical diversity in cultures can no longer be sustained by linguistics."[36]

So to return to the question—just what is it that is so much better in hunter-gatherer life? How does one encapsulate what can be sifted from an enormous body of scientific literature? It is not only, or even mainly, a matter of how nature is perceived, but of the whole of personal existence, from birth through death, among what history arrogantly calls "pre-agricultural" peoples. In the bosom of family and society, the life cycle is punctuated by formal, social recognition with its metaphors in the terrain and the plant and animal life. Group size is ideal for human relationships, including vernacular roles for men and women without sexual exploitation.[37] The esteem gained in sharing and giving outweighs the advantages of hoarding. Health is good in terms of diet as well as social relationships.[38] Interpenetration with the nonhuman world is an extraordinary achievement of tools, intellectual sophistication, philosophy, and tradition. There is a quality of mind, a sort of venatic phenomenology. "In a world where diversity exceeds our mental capacity nothing is impossible in our capacity to become human."[39] Custom firmly and in mutual council modulates human frailty and crime. Organized war and the hounding of nature do not exist. Ecological affinities are stable and nonpolluting. Humankind is in the humble position of being small in number, sensitive to the seasons, comfortable as one species in many, with an admirable humility toward the universe. No hunter on record has bragged that he was captain of his soul. Hunting, both in an evolutionary sense and individually, is "the source of those saving instincts that tell us that we have a responsibility towards the living world."[40]

To make such statements is to set out the game board for the dialectics of our intellectual life. Graduate students, religious fundamentalists, economists, corporate executives, and numerous others, including a gleeful band of book reviewers, will leap to prove differently. I have a wonderful set of newspaper book reviews of *The Tender Carnivore* with headings like "Professor Says Back to the Cave" and "Aw, Shoot!" And there is always an anthropologist somewhere to point to a tribe that is an exception to one or another of the "typical" characteristics of hunter-gatherers, hence there can be no "universals," and so on.

The most erudite essay on hunting, ancient or modern, is Ortega y Gasset's *Meditations on Hunting*. He conceives the hunt in terms of "authenticity," especially in its direct dealing with the inescapable and formidable necessity of killing, a reality faced in the "generic" way of being human. He also refers to the hunter's ability to "be inside" the countryside, by which he means the natural system—"wind, light, temperature, round-relief, minerals, vegetation, all play a part; they are not simply there, as they are for the tourist or the botanist, but rather they function, they act." Ultimately, this function is the reciprocity of life and death. The enigma of death and that of the animal are the same, and therefore "we must seek his company" in the "subtle rite of the hunt." In all other kinds of landscape, he says—the field, grove, city, battleground—we see "man traveling within himself," outside the larger reality.

The humanized and domesticated places may have their own domestic reality, but Ortega refers to generic being. Ortega's is a larger understanding; he attends to human "species-specific" traits and escapes the cultural relativism and social reduction that have dominated anthropology. A biologist turned philosopher-historian, Ortega links "primitive" hunter-gatherers to ourselves. This is because there are characteristics of humankind, as Eibes-Eibesfeldt tells us,[41] as well as shared characteristics of hunter-gatherers, present and past.

What has been learned about the nature of our own problems in the past twenty years?

Item: Health disorders are increasingly traced to polluting poisons and to a diet of domesticated (i.e., chemically altered or chemically treated) plants and animals. More people every year eat the meat of wild animals, seek "organic" vegetables, and seek alternatives to chemicalized nature.

Item: Evidence indicates that the small, face-to-face, social group works better in the quality of social experience and decision-making for its members and in its efficacy as a functional institution.[42]

Item: Percussive music and great intervals of silence are evidently conducive to our well-being. A meditative stillness, suggests Gary Snyder, was invented by waiting hunters.[43] Perhaps this reflected the poised and ruminating hush of mothers of sleeping infants. High levels of sound have been directly linked to degenerative disease in urban life.

Item: Regular exercise, especially jogging, rare in 1965, was common by 1980. The sorts of exercise for men and women (aerobics, jogging, stretching) correlate with certain routines of life in cynegetic societies. The benefits are not only physical but mental.[44]

Item: One of the hardest stereotypes about the savage to die is gluttony. In arguing that Pleistocene peoples were responsible for the extinctions of large mammals, Paul Martin projected urban greed on the ancient hunters.[45] This preposterous theory ignores fundamental ecology, comparative ethnography, and the anthropological distinctions between people who maximize their take and those who optimize it.[46] Given the whole range of Pleistocene extinctions it is a poor fit in the paleontological and archaeological record.

Item: Childhood among hunter-gatherers better fits the human genome[47] in terms of the experience and satisfaction of both parents and children. I refer to the "epigenetic" calendar, which is based on the complex biological specialization of neoteny, to which human culture is in part mediator and mitigator.

Item: That advanced intelligence not only arrived with hunting and being hunted but continues to be the central characteristic of the hunt is still hard to accept for those who think of predation as something like a dogfight. Knowledge is of overwhelming importance in accommodating the whole of society to a "watchful world" and structuring the mentality of the hunter. There are three evolutionary correlates of large cerebral hemispheres: large size, predator-prey interaction, and intense sociality.[48]

Item: The cosmography of tribal peoples is as intricate as any, and marked by a humility that is lacking in civilized society. For example, two of the "principles of Koyukon worldview" are "each animal knows way more than you do," and "the physical environment is spiritual, conscious, and subject to rules of respectful behavior."[49] The essays in Gary Urton's *Animal Myths and Metaphors in South America*[50] describe myths of the sort depicted in Huichol yarn paintings of Mexico—visual evocations of stories that integrate the human and nonhuman in dazzling, sophisticated metaphor.

The Paradox of the Civilized Hunter

There is no room here to review current ideas about hunting by modern, urban people, except to observe that the argument for hunting links primitive and civilized people, past and present. One can split this distinction and say with Barry Lopez that hunting is okay for ethnic groups but not for modern people. I think that view is based mistakenly on the notion that there are vicarious alternatives and reflects a kind of despair over the practical question of how the sheer numbers of people now living could gain the benefits of hunting-gathering.

Antihunters are outraged by "sport killing" as opposed to ethnic tradition, pointing for example to the diminished presence of wildlife and to old photographs of white African hunters with numerous dead animals.

Who would consider defending such "slaughter"? What is sometimes regarded as vanity needs to be understood in the context of the traditional laying out of the dead animals. One of the most thoughtful modern hunters, C. H. D. Clarke, writes, "The Mexican Indian shamanic deer hunt is as much pure sport as mine, and the parallels between its rituals, where the dead game is laid out in state, and those of European hunts, where the horns sound the 'Sorbiati,' or 'tears of the stag,' over the dead quarry, are beyond coincidence."[51]

Fanatic opposition to hunting suggests that some other fear is at work. Neither the animal protectionists, the animal rights philosophers, nor the feminists hostile to vernacular gender have ecosystems (including the wildness of humans) at heart. When antihunters heard that "a Royal Commission on blood sports in Britain reported that deer had to be controlled and that hunting was just as humane as any alternative, these people wanted deer exterminated once and for all, as the only way to deliver the land from the infamy of hunting." In America we have similar ecological blindness regarding the killing of goats on the coastal islands of California and wild horses in national parks. I once heard a nationally known radio commentator, Paul Harvey, complain that the trouble with the idea of national parks protecting both predators and prey animals was that "mercy" was missing. Clarke concludes that the "rejection of hunting is just one in a long list of rejections of things natural," and that hunting will linger as one of the human connections to the natural environment "until the human race has completed its flight from nature, and set the scene for its own destruction."[52]

Romancing the Potato

Seventeen years after the publication of *The Tender Carnivore* there is still only speculation among scholars about the "cause" of the first agriculture. It is clear now as it was then, however, that recent hunting-gathering peoples did not joyfully leap into farming. The hunter-gatherers' progressive collapse by invasion from the outside is typified in Woodburn's description of the Haida.[53] For ten millennia there has been organized aggression against hunters, who themselves had no tradition of war or organized armies. The psychology of such assault probably grew out of the territoriality inherent in agriculture and farmers' exclusionary attitude toward outsiders, land hunger growing from the decline of field fertility and the increase in human density, and, with the rise of "archaic high civilizations," social pathologies related to group stresses and insecurity in an economy of monocultures (i.e., grains, goats), and the loss of autonomy in the pyramiding of power. Hunting-gathering peoples have been the victims of these pressures that beset farmers and ranchers, bureaucratically amplified upward in the levels of government.

The old idea that farming favored more security, longer life, and greater productivity is not always correct. For example, Marek Zvelebil, in *Scientific American* in 1986, says, "Hunting-and-gathering is often thought of as little more than the prelude to agriculture. A reevaluation suggests it was a parallel development that was as productive as early farming in some areas."[54] As for modern agriculture, C. Dean Freudenberger says, "Agriculture, closely related to global deforestation by making room for expanding cropping systems, is the most environmentally abusive activity perpetuated by the human species.[55]

At least six millennia of mixed tending and foraging followed the first domesticated wheat and preceded the first wheel, writing, sewers, and armies. In varying degrees local, regenerative, subsistence economies blended the cultivated and gathered, the kept animal and the hunted. Before cities, the world remained rich, fresh, and partly wild beyond the little gardens and goat pens. Extended family, small-scale life with profound incorporation into the rhythms of the world made this "hamlet society" the best life humans ever lived in the eyes of many. It is this village society of horticulture, relatively free of monetary commerce and outside control, that most idealizers of the farm look to as a model.

Perhaps that image motivated Liberty Hyde Bailey in his turn-of-the-century book, *The Holy Earth*. Yet his feeling for the land seems betrayed by a drive to dominate. Bailey says, "Man now begins to measure himself against nature also, and he begins to see that herein shall lie his greatest conquests beyond himself; in fact, by this means shall he conquer himself—by great feats of engineering, by complete utilization of the possibilities of the planet, by vast discoveries in the unknown, and by the final enlargement of the soul; and in these fields shall he be the heroes. The most virile and upstanding qualities can find expression in the conquest of the earth. In the contest with the planet every man may feel himself grow."[56] Tethering the neolithic reciprocity with a nourishing earth, he suddenly jerks us into the heroic Iron Age. In the same book, however, he says, "I hope that some reaches of the sea may never be sailed, that some swamps may never be drained, that some mountain peaks may never be scaled, that some forests may never be harvested."[57] Inconsistent? No, it is an expression of the enclave mentality, the same one that gave us national parks and Indian reservations, the same that gives us wilderness areas.

The ideal of hamlet-centered life is represented by *Mother Earth News*, a search for equilibrium between autonomy and compromise. It is difficult not to be sympathetic. So too do Wes Jackson and the "permaculture" people seem to seek the hamlet life.[58] Their objective of replacing the annual plants with perennials seems laudable enough. Yet they are busily domesticating through selective breeding more wild perennials as fast as possible. They are making what geneticist Helen Spurway called genetic "goofies," the tragic deprivation of wildness from wild things.[59]

Who among us is not touched by the idyll of the family farm, the Jeffersonian yeoman, the placeness and playground of a rural existence? Above all, this way of life seems to have what hunting-gathering does not—retrievability. The yearning for it is not from academic studies of exotic tribal peoples, but is only a generation or two away—indeed, only a few miles away in bits of the countryside in Europe and America. After all, it incorporates part-time hunting and gathering, as though creating the best of all possible worlds. Like many others, I admire Jefferson as the complete man and share the search for peace of mind and good life of modern spokespeople like Wendell Berry.

Of course, most agriculture of the past five millennia has not been like that. The theocratic agricultural states, from the early centralized farms in farms ancient Sumer onward, have been enslaving rather than liberating. Even where the small scale seems to prevail, such conviviality is not typical in medieval or modern peasant life, with its drudgery, meanness, and suffering at the hands of exploitive classes above it.[60]

The primary feature of the farmer's concept of reality is the notion of "limited good." There is seldom enough of anything. By contrast, the hunter's world is more often rich in signs that guide toward a gifting destiny in a realm of alternatives and generous subsistence. Since they know nature well enough to appreciate how little they know of its enormous complexity, hunter-gatherers are engaged in a vast play of adventitious risk, hypostatized in gambling, a major leisure-time activity. Their myths are rich in the strangeness of life, its unexpected boons and encounters, its unanticipated penalties and mysterious rewards, not as arbitrary features but as enduring, infinitely complex structure. Gathering and hunting are a great, complex cosmology in which a numinous reality is mediated by wild animals. It is a zero-sum game, a matter of leaning toward harmony in a system which they disturb so little that its interspecies parities seem more influenced by intuition and rites than physical actions. Autonomous, subsistence farming or gardening shares much of this natural reverence for the biotic community and the satisfactions of light work schedules, hands-on routines, and sensitivity to seasonal cycles.

But agriculture, ancient and modern, is increasingly faced with a matter of winners and losers, dependence on single crops. Harmony with the world is sustained by enlarging the scope of human physical control or by rites of negotiation with sacred powers, such as sacrifice. The domesticated world reduces the immediate life forms of interest to a fewscore species that are dependent on human cultivation and care—just as the farmers see themselves, dependent on a master with humanlike, often perverse actions. Theirs is a cosmos controlled by powers more or less like themselves, from local bureaucrats up through greedy princes to jealous gods. No wonder they prefer games of strategy and folktales in which the "animals," burlesques of their various persecutors, are outwitted by clever foxes like themselves. The world does not so much have parts as it has sides substructured as class. From simple to complex agriculture these increase in importance as kin connections diminish.

The transition from a relatively free, diverse, gentle subsistence to suppressed peasantry yoked to the metropole is a matter of record. The subsistence people clearly long for genuine contact with the nonhuman world, independence from the market, and the basic satisfaction of a livelihood gained by their own hands. But this distinction among agricultures has its limits and was not apparently in mind when Chief Washakie of the Shoshones said, "God damn a potato." Sooner or later you get just what the Irish got after they thought they had rediscovered Eden in a spud skin.

We may ask whether there are not hidden imperatives in the books of Wendell Berry obscured by the portrayal of the moral quality, stewardship syndrome, and natural satisfactions of farm life. He seems to make the garden and barnyard equivalent to morality and aesthetics and to relate it to monotheism and sexual monogamy, as though conjugal loyalty, husbandry, and a metaphysical principle were all one. And he is right. This identity of the woman with the land is the agricultural monument, where the environment is genderized and she becomes the means of productivity, reciprocity, and access to Otherness, compressed in the central symbol of the goddess. When the subsistence base erodes, this morality changes. Fanaticism about virginity, women as pawns in games of power, and their control by men as the touchstone of honor and vengeance have been clearly shown to be the destiny of subequatorial and Mediterranean agri-culture.[61] Aldous Huxley's scorn of Momism is not popular today, but there are reasons to wonder whether the metaphors that mirror agri-culture are not infantile.[62] (For hunter-gatherers the living metaphor is other species, for farmers it is mother, for pastoralists the father, for urban peoples it has become the machine.)[63]

In time, events and people seem to come back in new guise. I keep thinking that Wendell Berry is the second half-century's Louis Bromfield. Bromfield was a celebrated author and gentleman farmer, known for his conservation practices and the good life on his Ohio farm. He could prove the economic benefit of modern farming by his detailed ledgers. But it was his novels that made him wealthy, and the dirt farmers who were invited along with the celebrities to see his showplace could well ask, "Does Brom-field keep books or do the books keep Bromfield?"

Berry writes with great feeling about fresh air and water, good soil, the sky, the rhythms of the earth, and human sense in these things. But those were not invented by farmers. They are the heritage of the nondomesti-cated world. Much that is "good" in his descriptions does not derive from its husbandry but from the residual "wild" nature. He accepts biblical admon-ishments about being God's steward, responsible for the care of the earth. None of the six definitions of "steward" in my dictionary mentions respon-sibility toward what is managed. It refers to one who administers another's property, especially one in charge of the provisions—another way of saying that the world biomes need to be ruled, that nature's order must be imposed from the outside.

Alternatively, one could pick any number of Christian bluenoses, from popes to puritans and apostles to saints, who wanted nothing to do with nature and who were disgusted to think they were part of it. The best that can be said about Christianity from an ecological viewpoint is that the Roman church, in its evangelical lust for souls, is a leaky ship. Locally it can allow reconciliation of its own dogma with "pagan" cults, as when the Yucatán Indians were Christianized by permitting the continued worship of limestone sinks, or *cenotes*, making the Church truly catholic.[64] Similar blending may be seen in eccentrics like St. Francis or Wendell Berry, who voice a "tradition" that never existed.

The worst is difficult to choose, although its shadow may be discerned behind the figure of Berry himself in *The Unsettling of America*, humming his bucolic paeans to the land and clouds and birds as he sits astride a horse, his feet off the ground, on that domestic animal which more than any other symbolized and energized the worldwide pastoral debacle of the skinning of the earth, and the pastoralists' ideology of human dissociation from the earthbound realm. No wonder the horse is the end-of-the-world mount of Vishnu and Christ. As famine, death, and pestilence, it was the apocalyptic beast who carried Middle East sky-worship and the sword to thousands of hapless tribal peoples and farmers from India to Mexico.

Dealing with Death

Joseph Campbell, who clearly understood the hunter-gatherer life, tried to have it both ways. The hunters' rituals, he said (capitulating to the nineteenth-century anthropological opinion that primitive religion is simply bad logic), tried to deny death by the pretense that a soul lived on. "But in the planting societies a new insight or solution was opened by the lesson of the plant world itself, which is linked somehow to the moon, which also dies and is resurrected and moreover influences, in some mysterious way still unknown, the lunar cycle of the womb."[65] The planters did indeed lock themselves to the fecundity and fate of annual grains (and their women to an annual pregnancy). But according to Alexander Marshack the moon's periodicity had long since been observed by hunters. In any case it was not seen by the early planters of the Near East as a plant but as a bull eaten by the lion sun.

Campbell regards sacrifice as the central rite of agriculture's big idea that the grain crop is the soul's metaphor. Sacrifice—the offering of fruit or grain, or the ritual slaughter of an animal or person—is a means of participating in the great round. But in agriculture participation turns into manipulation. The game changes from one of chance to one of strategy, from reading one's state of grace in terms of the hunt to bartering for it, from finding to making, from a sacrament received to a negotiator with anthropomorphic deities. This transition can be seen in a series of North Asian forms of the ceremony of the slain bear, from an egalitarian, ad hoc though traditional, celebration of the wild kill as a symbolic acceptance of the given to the shaman-centered spectacle of the sacrifice of a captive bear to deflect evil from the village.[66] The transition from bear hunt to bull slaughter has been traced by Tim Ingold.[67] Sacrifice does not seem to me to accommodate the "problem of death" but to domesticate it. It reverses the gift flow idea from receiving according to one's state of grace to bartering, from the animal example of "giving away" to the animal's blood as currency.

The changes that take place as people are forced from hunting-gathering to agriculture are not conjectural, but observed in recent times among the !Kung.[68] Their small-group egalitarian life vanishes beneath chiefdoms, children become excessively attached and more aggressive, and there are more contagious diseases, poorer nourishment, more high blood pressure, earlier menarche, three times as many childbirths per woman, and a loss of freedom in every aspect of their lives.[69] The farmer remains lean if he is hungry, but otherwise his body loses its suppleness. One might well wonder who benefits from all this, and of course the answer is the landholders, middlemen, bureaucrats, white-collar workers, and corporations. It is their spokespeople who echo C. H. Brown's blithe view that "a major benefit of agriculture is that it supports population densities many times greater than those that can be maintained by a foraging way of life." He adds, "Of course, this benefit becomes a liability if broad crop failure occurs."[70] He does not say who benefits from the bigger population density, and he is wrong about the "if" of crop failure—it is only a matter of "when."

Today most of us live in cities, but the leftover ideology of farming is the basis—ever since the Greek pastoral poets, Roman bucolics, and later the European rustic artists—of the nature fantasies of urban dwellers. Its images of a happy yeomanry and happy countryside are therapeutic to the abrasions of city life. This potato romance is not only one of celebrating humanity surrounded by genetic slaves and freaks but of perceiving the vegetable world as a better metaphor. The heritable deformity of cows and dogs is inescapable, while carrots and cereal grains seem fresh from the pristine hand of nature. This post-Neolithic dream lends itself, for example, to the recovery of the paradisiacal ecological relations of a no-meat diet.

The Vegetarians

The ethical-nutritional vegetarians, the zucchini-killers and drinkers of the dark blood of innocent soybeans, argue for quantity instead of quality. The Animal Aid Society's "Campaign to Promote the Vegetarian Diet" calculates that ten acres will feed two people keeping cattle, ten eating maize, twenty-four munching wheat, and sixty-one gulping soya.[71] The same space would probably support one or fewer hunter-gatherers. There is nothing wrong with their humane effort "toward fighting hunger in the Third World" of course, but what is life to be like for the sixty-one people and what do we do when there are 122 or 488? And what becomes of the Fourth World of tribal peoples or the Fifth World of nonhuman life?

The quantitative-mindedness links them philosophically with the nationalistic maximizers who assume that military advantage belongs to the most populous countries, with the politics of growth-economists and with the local greed for sales. Nutritionally, energy increase is no substitute for protein quality, nor adipose fats for the structural fats necessary for growth and repair, nor calories over immune system needs, or over the proportions of vitamins and essential minerals found in animal tissues.

Apart from their demographic and ecological short-sightedness, the vegetarians rightfully reject the fat-assed arrogance of piggish beefsteak-eaters, but they become slaves to protein hunger by striving to get eight of the twenty amino acids that their own bodies cannot make and that meat contains in optimum amounts. The search leads to cereals and legumes; the first are low in lysine, the second in methionine. Humans with little or no meat must get combinations of legumes and grain (lentils and rice, rice and beans, corn and beans), and they must locate a substitute source for vitamin B-12, which comes from meat.

Just this side of the vegetarians are various degrees of meat eating, and the same chains of reasoning carry us from red to white meats and from meat to eggs and milk. Neither domestic cereals nor milk from hoofed animals are "natural" foods in an evolutionary sense; witness the high levels of immune reaction, cholesterol susceptibility, and the dietary complications from too much or too little milling of grains.

Except for a tiny minority, people everywhere, including farmers, prefer to eat meat, even when its quality has been reduced by domestication. Marvin Harris has summed up the evidence from ethnology and physiology: "Despite recent findings which link the overconsumption of animal fats and cholesterol to degenerative diseases in affluent societies, animal foods are more critical for sound nutrition than plant foods."[72]

Nutritionally, little detailed comparison has been made between domestic and wild meats. Long-chain fatty acids, found only in meat, are necessary for brain development. These come from structural rather than adipose fat. You can get them in meat from the butcher, but domestic cattle often lack access to an adequate variety of seeds and leaves to make an optimum proportion of structural fats.[73] The latter are richest in wild meats.

Theories that attempt to center human evolution around something like the role of female chimpanzees, or to link gathering with a gender-facilitated evolution by reference to the "vegetarian" diets of primates, neglect the protein-hunger of primates and their uptake of meat in insect and other animal materials. The argument that humans are physiologically "closer" to herbivory than to carnivory, somehow placing women closer to the center of human being, is a red herring based on a mistaken dichotomy. It simply ignores human omnivory, signified not only in food preferences but physiologically in the passage time of food in the gut (longer in herbivores because of the slow digestion of cellulose-rich and fibrous foods, shorter in carnivores). In humans it is half-length between gorillas and lions.

Among most tribal peoples meat comprises less than 50 percent of the total diet most of the time, the bulk being made up of a wide variety of fruits and vegetables. But meat is always the "relish" that makes the meal worthwhile, and close attention is always paid to the way meat is butchered and shared. Vegetarianism, like creationism, simply reinvents human biology to suit an ideology. There is no phylogenetic felicity in it.

As for the alternatives in turning from the cholesterol of domestic meats, not everything comes up yogurt. Many European restaurants now offer a separate menu of game animals (reared but not domesticated). S. Boyd Eaton and Marjorie Shostak, an M.D. and an anthropologist, comment: "The difference between our diet and that of our hunter-gatherer forebears may hold keys to many of our current health problems. . . . If there is a diet natural to our human makeup, one to which our genes are still best suited, this is it."[74]

Cultural Evolution

The casual misuse of "evolution" in describing social change produced enough confusion to mislead generations of students. Every society was said to be evolving somewhere in a great chain of progress. Beginning in a Heart of Darkness in the individual and at the center of remote forests humankind advanced to ethics, democracy, morality, art, and the other benefits of civilization. This ladder probably still represents the concept of the past for most modern, educated people. It is a direct heritage of the Enlightenment and its industrial science, its spectatorship (as in the art museum or at the play), elitism, and the cult of the *polis*.

Recently there have appeared new versions of lifeways that refute a universal yearning toward civilization, from savagery through nomadic pastoralism and various agricultures to a pinnacle of urban existence.[75] The revised version also denies a hierarchy of inherent physical or mental differences among the peoples of different economies.

One modified view presents us with shifts in which societies are compelled to change not so much as an advance as a result of circumstances beyond their control—increased population density and the struggle for power and space. It offers a "circumscription theory." Societies at the denser demographic end show a hierarchical, imperial domain and the loss of local autonomy in which symbols of participation in the larger system replace real participation for the individual. Such societies subjugate or are conquered by others.

In a recent book Allen Johnson and Timothy Earle cite specific examples from first to last.[76] They begin with a description of hunting-gathering at the family level of economy, characterizing it as low in population density, making personal tools, engaged in annual rhythms of social aggregation and dispersion, informally organized with ad hoc leadership, collectively hunting large game, lightly assuming tasks of gathering, without territoriality or war, and with numerous alternatives in "managing risk."

Such easy going societies continue with minor introduction of domestic plants and animals, at the same time consciously resisting life in denser structures. In villages, however, men begin to fight over "the means of reproduction" and depart from the "modesty and conviviality" found in family-level societies. As "geographical circumscription" closes around them, leaving nowhere to go, there is more bullying, impulsive aggression, revenge, and territoriality. "Scarcity of key resources" and war become "a threat to the daily lives" of these horticulturalists and pig-raisers. As the economy "evolves" the "domestication of people into interdependent social groups and the growth of political economy are thus closely tied to competition, warfare and the necessity of group defense."

As villages get bigger, Johnson and Earle continue, "Big Man" power appears, ceremonial life shifts from cosmos-focused family activity to public affirmation of political rank. Dams and weirs and slaves and food surplus and shortage management occupy the leaders. But "the primary cause of organization elaboration appears to be defensive needs." Among typical yam-growers of the South Pacific, "half a mile beyond a person's home lies an alien world fraught with sudden death."

Meanwhile, the pastoralists also "evolve." Their lives are increasingly centralized under patriarchal systems based on "friends" who "help spread the risk" of resource depletion and defense needs. As cattle become currency, raiding and banditry increase in a "highly unpredictable environment." Chiefdoms are subordinated by greater chiefs, who allocate pasture and travel lanes, manage "disagreement resolution" locally, and negotiate alliances and conflict externally. Life is lived in camp, that is, "a small nucleus of human warmth surrounded by evil." Their equivalents in sedentary towns are concerned with crop monocultures and massive tasks of "governing redistribution," regulating the bureaucracy and management of field use and irrigation works.

When we get to the first true or archaic states, vassalage, standing armies, and taxes make their appearance. "Social circumscription" is added to geographical circumscription. Religion and staple-food storage is centralized. As the state matures, the peasants emerge with "no end of disagreement and even disparagement" among themselves. They often "live so close to the margin of survival that they visibly lose weight in the months before harvest." As we approach the modern state, the authors say, "peasant economics provide a less satisfactory subsistence than the others we have examined," with poor diet, undernourishment, extreme competition, and a meager security experienced as vulnerability to markets controlled from the outside or the arbitrary will of patrons.

Johnson and Earle conclude, at the end of this long road to a "regional polity," that the record is one of endless rounds of population increase and "intensification," producing societies symbolized by their dependence on "starchy staples." All hail the potato.

The authors are careful to remain mere observers. If a book can have a straight face while taking off civilization's pants, here is a wonderful irony, although probably a competent synthesis of the record. Yet euphemisms and semitechnical phrases abound. For "diminished resources" one should read "collapse of life support" or "failed ecosystems." For "local slave management" read "tyranny," for "risk management" simply "debacle." The increasing need for "defense" is frequently mentioned, but who is doing all the offense? How casually and with value-free candor we move from many options in "risk management" to few, from personal tools to work schedules, from ad hoc leadership to hierarchies of chiefs. Little is said about children, women, the source of slaves, the loss of forests and soil, the scale of tensions between farmers and pastoralists. One has to interpolate the relevant changes in the role and status of women, the lives of children, or the condition of the nonhuman fellow-beings. The book seems to achieve its objective of combining "economic anthropology and cultural ecology," making disaster humdrum and so inevitable. The recitation of the "evolution of culture" in such expressionless fashion is in fact enormously effective, for the authors seem oblivious to the horrors they describe. I am reminded of academics who reply to descriptions of the biotic costs of civilization with murmurings about how difficult life would be for them without Beethoven, cathedrals, and jurisprudence. But then, it was a tiny elite who benefited from this "evolution" all along, and I suppose that they can easily imagine that others, in their benighted state, cannot possibly appreciate the gains.

For twenty years my students and colleagues have responded to this scenario by asking why people changed if the old way was better, and then refuse to believe that the majority were compelled by centralized force in which power and privilege motivated the few. Zvelebil says, "The stubborn persistence of foraging long after it 'should' have disappeared is one of the qualities that is contributing to a fundamental reassessment of postglacial hunting and gathering."

The idea of cultural change as a paradoxical "development" can also be seen in a comparison of American Indian tribes. John Berry and Robert Annis studied differences in six northern Indian tribes using George Murdock's classifications of culture types, "a broad ecological dimension running from agricultural and pastoral interactions with the environment through to hunting and gathering interactions." They describe a corresponding psychological differentiation, defined along this axis.

Agriculture tends to be associated with high food accumulation, population density, social stratification, and compliance. At the other end of the series are the low food accumulators—hunter-gatherers—with a high sense of personal identity, social independence, emphasis on assertion and self-reliance, high self-control, and low social stratification. Berry and Annis see these differences in terms of "cognitive style," "affective style," and "perceptual style." These studies are consistent with the work of

Robert Edgerton, who found distinct personality differences between farmers and pastoralists.[78]

What we come to is an uneasy sense of economic determinism. There is a profound similarity among hunter-gatherers everywhere. This convergence demonstrates the nichelike effect of a way of life. The possibilities for human cultural mixtures can be seen in the variety of peoples in the modern world. There seems to be no end to the anthropological exploration of their differences. Still, the surprising thing is not their dissimilarity but the extent of common style. Something enormously powerful binds living hunter-gatherers to those of the past and to modern sportsmen.

They are all engaged in a game of chance amid heterogeneous, exemplary powers rather than in collective strategies of accumulation and control. Their metaphysics conceives a living, sentient, and dispersed comity whose main features are given in narrations that are outside History. Their mood is assent. Their lives are committed to the understanding of a vast semiosis, presented to them on every hand, in which they are not only readers but participants. The hunt becomes a kind of search gestalt. The lifelong test and theme is "learning to give away" what was a gift received in the first place.

There are also convergent likenesses among subsistence farmers, pastoralists, and urban peoples. The economic constraints seem to transcend religions and ethnic differences, to surpass the unique effects of history, to overstep ideology and technology. The philosophies as well as the material cultures of otherwise distant peoples who have similar ecologies seem to converge.

Wilderness and Wildness

Wilderness

How are we to translate the question of the hunt into the present? One road leads to the idea of wilderness, the sanctuaries or sacrosanct processes of nature preserved.

The idea of wilderness—both as a realm of purification outside civilization and as a place of beneficial qualities—has strong antecedents in the Western world. In spite of the recent national policies of designating wilderness areas, the idea of solace, naturalness, nearness to fundamental metaphysical forces, escape from cities, access to ruminative solitude, and locus of test, trial, and special visions—all these extend biblical traditions. As for wildness, I suppose that most people today would say that wilderness is where wildness is, or that wildness is an aspect of the wilderness.

Wilderness is a place you go for a while, an escape to or from. It is a departure into a kind of therapeutic land management, a release from our crowded and overbuilt environment, an aesthetic balm, healing to those who sense the presence of the disease but who may have confused its cause with the absence of the therapy. More important, we describe it to ourselves in a language invented by art critics, and we take souvenirs of our experience home as photographs. Typically, the lovers of wilderness surround themselves with pictures of mountains or forests or swamps which need not be named or even known, for they are types of scenery. But it is emphatically not scenery that is involved in either the ceremonies of aborigines or the experience of the hermit saints. Something has intervened between them and the *zeitgeist* of the calendar picture. That something is the invention of landscape.

Wilderness remains for me a problematic theme, intimately associated in the modern mind with landscape. It is a scene through which spectators pass as they would the galleries of a museum. Art historians attribute the origins of landscape (in the Occident) to sixteenth-century perspective painters, but I find a strange analogy to the descriptions of Mesolithic art, where "we are evidently approaching a historical sense. . . . The tiny size of these paintings is something of a shock after the Paleolithic. The immediate impression is of something happening at a great distance, watched from a vantage point which may be a little above the scene of the action. This weakens the viewer's sense of participating in what is going forward. There is something of a paradox here, for in the graphic art of the Paleolithic, though man was seldom shown, he was the invisible participant in everything portrayed, while now that he has moved into the canvas and become a principal, there is a quite new detachment and objectivity about his portrayal."[79] In other words, the first appearance of genre and perspective in pictorial art is Neolithic and probably expresses a new sense of being outside nature. Something like modern landscape reappears later in Roman mosaics, prior to its rediscovery by Renaissance art, and I take this as evidence of renewed "distancing" and an expression of the Classical rationality that made possible the straight roads across Europe, based on survey rather than old trails.

I owe to David Lowenthal and Marshall McLuhan a debt for diverting me from writing and thinking about wilderness. Graduate work on the history of landscape, published as *Man in the Landscape,* left me susceptible to McLuhan's devastating analysis of seventeenth-century science and art. Linear/mathematical thinking and the representation of places as aesthetic objects distanced the observer from rather than connected him to his surroundings.[80] The place was framed. This was the aesthetic origin of pictorial vision, of which wilderness is a subject matter.

Lowenthal did not describe so much as embody the humanist position, in which the "love of nature" is understood as an aesthetic experience, and any aesthetic is a "congeries of feelings," a cultural ripple that can come and go in the dynamics of taste and fashion.[81] Lowenthal is wrong. He misunderstands the truly radical aspect of romanticism, misconstruing it as aesthetic or iconographic rather than an effort to reintegrate cognition and feeling in an organic paradigm. But he may be right about landscape. It was the means of perceiving nature according to criteria established by art criticism, the avenue of "landscape" by which people "entered" nature as they did a picture gallery. As long as pictures were regarded as representations, the enthusiasm for landscape could still penetrate all areas of culture, in spite of the estrangement described by McLuhan. By the end of the nineteenth-century the art world moved on to nonobjectivity, leaving wilderness with the obsolescence and superficiality with which Lowenthal confused it.

The landscape cannot escape its origins as an objectifying perception, although it may be misused as a synonym for place, terrain, ecosystem, or environment. Photos of it are surrealistic in the sense that they empty the subject of intimate context. As pictures age they add layers of a cold impulse like growing crystals, making the subject increasingly abstract, subjecting real events to a drifting, decadent attention. When nineteenth-century painters discovered photography they were freed, as Cezanne said, from literature and subject matter. Susan Sontag has it right about surrealism: disengagement and estrangement. It is, she says, a separation that enables us to examine dispassionately old photographs of suffering people.[82] It is a form of schizophrenia, a final effect of splitting art from its origins in religion. It becomes seeing for its own sake, what Bertram Lewin has called "neurotic scopophilia."[83] To this I add the photography of nature, which antihunters want to substitute for killing and eating. Pictures of nature exactly embody what is meant by wilderness as opposed to that wildness which I kill and eat because I, too, am wild.

Wildness

Thank God Thoreau did not say, "In wilderness is the preservation of the world." Wildness, ever since Starker Leopold's research on heritable wildness in wild turkeys in the mid-1940s and Helen Spurway's "The Causes of Domestication,"[84] has for me an objective reality, or at least a degree of independence from arbitrary definitions.

Wildness occurs in many places. It includes not only eagles and moose and their environments but house sparrows, cockroaches, and probably human beings—any species whose sexual assortment and genealogy are not controlled by human design. Spurway, Konrad Lorenz's observation on the bodily and behavioral forms of domesticated animals, and the genetics of zoo animals provide substance to the concept. The loss of wildness that results in the heritable, blunted, monstrous surrogates for species, so misleading because the plants and animals that seem to be there have gone, is like sanity's mask in the benign visage of a demented friend.

What then is the wild human? Who is it? Savages? Why . . . it is us! says Claude Lévi-Strauss. The savage mind is our mind.[85] Along with our admirable companions and fellow omnivores—the brown rat, raccoon, and crow, not yet deprived of the elegance of native biology by breeding management—it is us! Some among us may be deformed by our circumstances, like obese raccoons or crowded rats, but as a species we have in us the call of the wild.

It is a call corrupted not only by domestication but by the conventions of nature aesthetics. The corporate world would destroy wildness in a trade for wilderness. Its intent is to restrict the play of free and selfish genes, to establish a dichotomy of places, to banish wild forms to enclaves where they may be encountered by audiences, while the business of domesticating the planet proceeds. The savage DNA will be isolated and protected as aesthetic relics, as are the vestiges of tribal peoples. This includes the religious insights of wild cultures, whose social organization represents exotic or vestigial stages in "our" history or "evolution," their ecological relations translated into museum specimens of primeval economics. My wildness according to this agenda is to be experienced on a reservation called a wilderness, where I can externalize it and look at it.

Instead my wildness should be experienced in the growing of a self that incorporates my identity in places. See Fred Myer, Roy Rappaport, D. H. Stanner, or Gary Snyder on the way the self exists in resonance with specific events in particular places among Australian peoples.[86] The Australian

Outback is not a great two-dimensional space, not a landscape, but a pattern of connections, lived out by walking, ritually linking the individual in critical passages to sacred places and occasions, so that they become part of an old story. To be so engaged is like a hunger for meat, irreducible to starches, the wild aspect of ourselves.

Wild versus Domestic Metaphysics

The bones I sometimes think I have to pick with Gary Snyder are surely those remaining from a shared hunt and meal, pieces to be mulled over—to mull, from a root word meaning "to grind" or "to pulverize"—which I take to mean that we are sitting at a fire together, breaking femurs to get at the marrow or the pith.

He has said that the intent of American Indian spiritual practice is not cosmopolitan. "Its content perhaps is universal, but you must be a Hopi to follow the Hopi way," a dictum that all of us in the ragtag tribe of the "Wanta-bes" should remember. And he has said, "Otherworldly philosophies end up doing more damage to the planet (and human psyches) than the existential conditions they seek to transcend."[87] But he also refers to Jainism and Buddhism as models, putting his hand into the cosmopolitan fire, for surely they are two of those great, placeless, portable world religions whose ultimate concerns are not just universal but otherworldly. Yet without quite understanding why, from what I have seen of his personal life, there is no contradiction. I suspect that Snyder in the Sierra Nevada, like Berry in Kentucky and Wes Jackson in the Kansas prairie, is not so much following tradition but doing what Joseph Campbell called "creative mythology."

When I am sometimes discouraged by the thought that Gary Snyder has already said everything that needs to be said, as in, for instance, "Good, Wild, Sacred,"[88] I reawaken my independence of spirit by thinking of his faith in agriculture and Buddhism, even though in reality he carefully qualifies both. No matter how benign small-scale garden-horticulture may be, at its center is the degenerating process of domestication, the first form of genetic engineering. Domestication is the regulated alteration of the genomes of organisms, making them into slaves that cannot be liberated, like comatose patients hooked without reprieve to the economic machine.

As for coma, the excessive use of slave animals in experimental laboratories, their fecundating overspill as pets into city streets, and their debasement in factory farms has generated the "humane" movement, the dream of animal rights groups that by kindness or legislation you can liberate enslaved species. The clearest analogy is the self-satisfied, affectionate care of slaves by many pre–Civil War gentry. In our time, a huge, terrible yearning has come into the human heart for the Others, the animals who nurture us now as from our beginnings. Our gratitude to them is deep—so deep that it is subject to the pathologies of our crowded lives. In our wild hunger for the recovery of animal presence we have made and given names to pets, molded their being after our cultural emphasis on individuals. Our hunting past tells us that the species is the "individual," each animal the occasion of the species' soul. Our humane movement personalizes them instead, losing sight of the species and its ecology. Worse, that self-proclaimed "kindness" marks the collapse of a metaphor central to human consciousness, replacing it with the metonymy of touch-comfort, hence the new jargon of "animal companion" for pet in the new wave of "animal facilitated therapy." It is a massive, industrial effort among an amalgam of health workers, veterinarians, pet food manufacturers, and institutions. The effects of the therapy are undoubtedly genuine, but its "cognitive style" connects at one end with the hair-splitting philosophical rationality of the animal ethicists and at the other with the maudlin neuropath keeping thirty cats in a three-room city house—an abyssal chaos of purposes and priorities.

The lack of ecological concern in almost all animal ethics is strangely similar to that "embarrassed silence" in anthropology—the posture of detached respect by which all ethnic rites are interpreted as serving social and symbolic functions for an erroneous religion. Animal ethics comes from the same Greek source as all our philosophy, passionately reasoning but grounded in detachment and skepticism. There seems to be no real feeling there for the living world. They simply do not ask whether the Holy Hunt might indeed be so.

As for killing animals to eat, in *The Tender Carnivore* I suggested, taste buds and tongue in check, that in an overpopulated world we could free the animals, including ourselves, make hunting possible, and terminate the domestication of multicellular life by eating oil-sucking microbes (which is entirely feasible). To my surprise I find that this is our direction, in our yogurt and cheese rush to avoid killing "higher" animals by substituting down a chain of being, killing asparagus instead of cows or yeasts instead of asparagus. But there is no escape from the reality that life feeds by death-dealing (and its lesson in death-receiving). The way "out" of the dilemma is into it, a way pioneered for us in the play of sacred trophism, the gamble of sacramental gastronomy, central myths of gifts, and chance, the religious context of eating in which the rules are knowing the wild forms who are the game. You cannot sit out the game, but must personally play or hide from it.

This brings us back to Buddhism. I remain a skeptical outsider, unnerved by the works of Gary Snyder and Alan Watts, whose combined efforts I consider to be a possible library on how to live. Still, the Hindus disdained Buddhism when they discovered how abstract and imageless it was, how shorn of group ceremony, the guiding insights of gifted visionaries and the demonstrable respect for life forms represented in their multitudinous pantheon. The Hindus at least saw personal existence as a good many slices of *dharma* in a variety of species before the individual finally escaped into the absolute, while the Buddhists argued that all you needed was the right discipline and you could exit pronto.

The Buddhists' contemporaries and fellow travelers, the Jains, famous for *ahimsa* (harmlessness), are familiarly portrayed moving insects from the footpath. But this is not because they love life or nature. The Jains are revolted by participation in the living stream and want as little as possible to do with the organic bodies, which are like tar pits, trapping and suffocating the soul. Historically, it would appear that both Buddhists and Jains got something from the Aryans, who brought their high-flying, earth-escaping gods from Middle East pastoralism. In the face of these invasions, the Hindus and their unzipped polytheism survived best in the far south of India, where the Western monotheists penetrated least.

At a more practical level, everywhere the "world" religions have gone the sacred forests, springs, and other "places" and their wild inhabitants have vanished. The disappearance of respect for local earth-shrines is virtually a measure of the impact of the otherworldly beliefs. Can there be a world religion of bioregions, a universal philosophy of place, an inhabitation of planet Earth with plural, local autonomy?

Perception as the Dance of Congruity

René Dubos once observed that humans can adapt (via culture) to "starless skies, treeless avenues, shapeless buildings, tasteless bread, joyless celebrations, spiritualess pleasures—to a life without reverence for the past, love for the present, or poetical anticipations of the future. Yet it is questionable that man can retain his physical and mental health if he loses contact with the natural forces that have shaped his biological and mental nature."[89] But, unless these "forces" are the characteristics he mentions, what are they? His list is made up entirely of acts within a social and cultural milieu, by customary definition not "natural." Something "natural" looms behind all this, mediated by culture.

Dubos's statement is preceded by the observation that the human genetic makeup was stabilized 100,000 years ago. He quotes Lewis Mumford, "If man had originally inhabited a world as blankly uniform as a 'high rise' housing development, as featureless as a parking lot, as destitute of life as an automated factory, it is doubtful that he would have had a sufficiently varied experience to retain images, mold language or acquire ideas."[90]

What is this something natural necessary to become cultural? What is between culture and nature, betwixt the phenomenal or palpable world and the conceptual and ceremonial expressions of it? Connecting the cognition and the outer world is the event/structure, linking entity and environment. It is perception, the precognitive act, mostly unconscious, which directs attention, favors preferences, governs sensory emphasis, gives infrastructure. Lee and Ong's distinctions between an "acoustical event world" and the "hypervisual culture" is just such a prior mode, giving primordial design to experience, limiting but not formulating the concepts and enactments by which events are represented.[91] Phonetic alphabet, pictorial space, and Euclidean theory are not only ideas and formulas but frames supporting a kind of liminal foreknowledge of assumptions and inclinations.

Emphasis on perception does not mean that we shape our own worlds irrespective of a reality, or that one person's perceptual process is as reliable as another's. Perception is not another word for taste. In this, says Morris Berman, it transcends "the glaring blind spot of Buddhist philosophy."[92] Its truest expression "by test" (my criteria: quality of life; ecological integrity) in the world is the empirical effect of its contiguity. It is the process of the first steps of directed attention and vigilance. Perceptual habit is style in the sense that Margaret Mead once used the term, to mean a pattern of movement and sensitivity, the lively net of predisposition emerging from our early grounding, finally affecting every aspect of one's expressive life. In our wild aspect such unconscious presentations are centered in dance and narration, surrounded by innumerable and wonderfully varied moral and aesthetic presences. It presents us with an intuition of rich diversity, whose "forces" are purposeful and sentient. From Dubos's treeless avenues to Mumford's parking lot, it is not a view that is absent, or things or wilderness. It is a way of expecting and experiencing, encountering inhabitance by a vast congregation of Others unlike us, yet, like our deepest selves, wild.

The Mosaic

We must now close the circle to that sweeping, four-word dictum that is intended to close the door on access to the primitive: "You can't go back."

The Structural Dimension

The hereditary material is organized as a linked sequence of separable genes and chromosomes. This genome is a mosaic of harmonious but distinct entities. This structure makes possible the mutation of specific traits and the independent segregation of traits, the accumulation of multiple factors, and both the hiding and expression of genes.

The structure of the natural community, the ecosystem, is likewise an integrated whole composed of distinct species populations and their niches. The fundamental concept of modern biology is its primary characteristic as a composite of linked and harmonious but separable parts. The whole is neither the sum of its parts nor independent of any of them. As with genes, substitutions occur. A given species can be totally removed by extirpation or introduced into new communities. Witness for example the constitution of the prairie without the buffalo and the continuity of ecosystems after the successful introduction of the starling into North America.

Human culture, being genetically framed and ecologically adapted, is also an integrated conglomerate. Stories, dances, tools, and goods are sometimes completely lost from a society. At other times they move from culture to culture, sometimes trailing bits of the context from which they come, sometimes arriving rough-edged and isolate, but being assimilated, modified or not, as a part of the new whole.

There is a common characteristic of each of the above examples from the genome of the individual, the material or expressive culture of a people, and the tapestry of the natural environment. The specific entity involves both a distinct portability and a working embeddedness. The reality is more complex but the principle is true: the capacity for a part to be transferred. It is then part of a new whole. The rest of the totality adjusts, the organism accommodates, the niche system stretches or contracts, the culture is modified.

Societies and cultures are mosaics. They are componential. Their various elements, like genes and persons, can be disengaged from the whole. Contemporary life is in fact just such an accumulation representing elements of different ages and origins, some of which will disappear, as they entered, at different times than others. The phrase "You cannot go back" can only mean that you cannot recreate an identical totality, but it does not follow that you cannot incorporate components.

"You can't go back" is therefore a disguise for several assumptions, which in turn may hide ways of perceiving or preconstructing experience. One is the paradigm of uni-direction, the idea that time and circumstances are linear. Yet we "go back" with each cycle of the sun, each turning of the globe. Each new generation goes back to already existing genes, from which each individual comes forward in ontogeny, repeating the life cycle. While it is true that you may not run the ontogeny backwards, you cannot avoid its replays of an ancient genome, just as human embryology follows a pattern derived from an ancestral fish. Most of the "new" events in each individual life are new only within a certain genetic octave and only in their combinations. New genes do occur, but the tempo of their emergence is in the order of scores of thousands of years. The difference between the genomes of chimpanzees and humans is about 1 percent. Of the 146 amino acids of the Beta chain of blood hemoglobin, the gorilla differs from humans at one site, the pig at ten, the horse at twenty-six.[93]

A paradox is evident: newness yet sameness; repetition and novelty, past and present. Recall that the historical consciousness of the West rejects this as illusory ambiguity. The rejection is a characteristic perceptual habit. In tribal life, such matters of identity ambiguity are addressed ritually in the use of animal masks and mimetic dances, on the grounds that we are both animal and human, a matter "understood" by certain animal guides. Genes are not only "how-to" information but are mnemonic, that is, memories. Ceremonies recall. The reconciliation of our own polythectic zoological selfhood is inherent in our ritualized, sensuous assent of multiple truth. It denies the contradiction, abolishes the either-or dichotomy in the simultaneous multitude that we are. Our primitive legacy is the resolution of contradiction by affirmation of multiplicity, plurality, and change.

In advocating the "primitive" we seem to be asking someone to give up everything, or to sacrifice something: sophistication, technology, the lessons and gains of History, personal freedom, and so on. But some of these are not "gains" so much as universal possessions, reified by a culture that denies its deeper heritage. "Going back" seems to require that a society reconstruct itself totally, especially that it strip its modern economy and reengage in village agriculture or foraging, hence is judged to be functionally impossible. But that assumption misconstrues the true mosaic of both society and nature, which are composed of elements that are eminently dissectible, portable through time and space, and available.

You can go out or back to a culture, even if its peoples have vanished, to retrieve a mosaic component, just as you can transfer a species that has been regionally extirpated, or graft healthy skin to a burned spot from a healthy one. The argument that modern hunting-gathering societies are not identical to Paleolithic peoples is beside the point. It may be true that white, former Europeans cannot become Hopis or Kalahari bushmen or Magdalenian bison-hunters, but removable elements in those cultures can be recovered or recreated which fit the predilection of the human genome everywhere.

Three Important, Recoverable Components: The Affirmation of Death, Vernacular Gender, and Fulfilled Ontogeny

Our modern culture or "mosaic" is an otherworldly monotheism littered with the road kills of species. Road kills—such trivial death contrasts sharply with that other death in which circumspect humans kill animals to eat them as a way of worship.

This ancient, sacramental trophism is as fundamental to ourselves as to our ancestors and distant cousins. The great metaphysical discovery by the cynegetic world was cyclicity. It emerged in the context of the rites of death, both human and animal, as part of this flow. It is as old as the Neanderthal observation of hibernating bears as models of life given and recovered, and as new as Aldo Leopold's story "Odyssey" in *A Sand County Almanac* telling of an atom from a dead buffalo moving through the chain of photosynthesis, predation, decay, and mineralization. These concepts are about the nutritional value of meat in human metabolism as a reflection of a larger "metabolism," and about the gift of human consciousness in a sentient world in which food-giving symbolizes connectedness. Animals on the medicine wheel of the Plains Indians were said to be those that knew how to "give away." " 'Each dot I have made with my finger in the dirt is an animal,' said White Rabbit. 'There is no one of any of the animals in this world that can do without the next. Each whole tribe of animals is a Medicine Wheel, in that it is the One Mind. Each dot on the Great Wheel is a tribe of animals. And parts of these tribes must Give-Away in order that they all might grow. The animal tribes all know of this. It is only the tribes of People who are the ones who must learn it.' "[94]

William Arrowsmith, observing that in our time "we cannot abide the encounter with the 'other.' . . . We do not teach children Hamlet or Lear because we want to spare them the brush with death. . . . A classicist would call this disease *hybris*. . . . The opposite of *hybris* is *sophrosyne*. This means 'the skill of mortality.' "[95] It is the obverse side of the "giving away" coin, the way of momentarily being White Rabbit, reminding the human hunter that he too once was a prey and, in terms of the cosmic circling-back, still is.

The difficult question of interspecies ethics centers on death-dealing. Death is the great bugaboo. How we resent its connection to food—and to life—and repress the figure of the dying animal. Gary Snyder's reply: "All of nature is a gift-exchange, a potluck banquet, and there is no death that is not somebody's food, no life that is not somebody's death. Is this a flaw in the universe? A sign of the sullied condition of being? 'Nature red in tooth and claw'? Some people read it this way, leading to a disgust with self, with humanity, and with life itself. They are on the wrong fork of the path."

Joseph Campbell has argued rightly that death was a great metaphysical problem for hunters and concluded wrongly that it was solved by planters, with their sacrifices to forces governing the annual sprouting of grain. But it was control, not acquiescence to this great round, that the agriculturalists sought. In the Neolithic, says Wilhelm Dupré, 'The individual no longer stands as a whole vis-à-vis the life-community in the sense that the latter finds its realization through a total integration of the individual—as is the case by and large under the conditions of a gathering and hunting economy."[96]

"Hunters" is an appropriate term for a society in which meat, the best of foods, signifies the gift of life, the obtaining and preparation of which ritualizes the encounter of life and death, where the human kinship with animals is faced in its ambiguity, and the quest of all elusive things is experienced as the hunt's most emphatic metaphor.

Vernacular Gender

And so we bring to and from the mosaic of lifeways the hunt itself. Some feminists object that too much is made of it. But they misunderstand this killing of animals as an exercise of vanity, which they see as characteristic of patriarchy. They note that only a third of the diet is meat, the rest from plants, mostly gathered by women, as though there were a contest to see who really supports the society. In this they merely reverse the sexist view. Like so much of extremist feminism it is just a new "me first." They point out that in most hunting-gathering societies the women gather most of the food that is eaten. This view has the same myopia as that of the vegetarians—the tendency to quantify food value in calories. In any case they are wrong, as meat is so much higher in energy that the net energy gained from hunting is as great as that from gathering.[97]

While it is true that the large, dangerous mammals are usually hunted by men in hunting-gathering societies, it has never been claimed that women only pluck and men only kill. The centrality of meat, the sentient and spiritual beings from whom it comes, and the diverse activities in relationship to the movement of meat and the animal's numinous presence through the society entail a wide range of roles, many of which are genderized. Insofar as the animal eaten is available because it has learned "to give away," there is no more virtue in the actual chase or killing than the transformation of its skin into a garment, the burying of its bones, the drumming that sustains the dancers of the mythical hunt, or the dandling of infants in such a society as the story of the hunt is told.

Meat, says Konner, is only 30 percent of the !Kung diet, but it equals the nutritional value of the plant foods and produces 80 percent of the excitement, not only during the hunt, but in group life. The metaphysics of meat. The hunt itself is a continuum. From its first plan to its storied retelling, from the metaphors on food chains to prayers of apology, this carnivory takes nothing from woman, though it clarifies the very different meaning that different kinds of foods have in expressive culture. Broadly understood, the hunt refers to the larger quest for the way, the pursuit of meaning and contact with a sentient part of the environment, and the intuition that nature is a language. Hunting is a special case of gathering.

A critical dimension of the hunt is the confrontation with death and the incorporation of substance in new life, in all forms of sharing and giving away. Women are traditionally regarded as keepers of the mystery of death-as-the-genesis-of-life, hence the hunt is clearly connected with feminine secrets and powers, and we are not surprised to see Artemis and her other avatars, the archaic "Lady of the Beasts," and the Paleolithic female figurines in sanctuaries where the walls are painted with hunted game. More value is placed on men than women only as the hunt is perverted by sexism and war. Indeed, it is possible that sexism comes into being with the doting on fertility and fecundity in agriculture and the androgenous "reply" of nomadic, male-dominated societies of pastoralism.

Hunting has never excluded women, whose lives are as absorbed in the encounter with animals, alive and dead, as those of men. If in some societies the practices of vernacular gender tend more often to relegate to the men the pursuit of large, dangerous game, it relegates to the women the role of singing the spirit of the animal a welcome, and to them the discourse at the hearth where she is the host. Roles and duties are divided, but not to make inequality. Among the Sharanahua of South America, the women, being sometimes meat-hungry, send off the men to hunt and sing the hunters to their task. They are commonly believed to transform men into hunters. Janet Siskind says, "The social pressure of the special hunt, the line of women painted and waiting, makes young men try hard to succeed." Women also hunt. Gathering, like hunting, is a lighthearted affair done by both men and women. The stable sexual politics of the Sharanahua, "based on mutual social and economic dependence, allows for the open expression of hostility," a combination of solidarity and antagonism that "prevents the households from becoming tightly closed units."[98]

Martin Whyte, comparing "cultural features in terms of their evolutionary sequence," concluded that as civilization evolves, "Women tend to have less domestic authority, less independent solidarity with other women, more unequal sexual restrictions and perhaps receive more ritualized fear from men and have fewer property rights than is the case in the simpler cultures."[99]

All in all a far cry from the more strident views, whether of feminists, the obsolete social evolution of the neo-Marxists, or the flight from life of the humane animal protectionists. On the whole, plant foods are not shared as ceremoniously as meat. They do not signify the flow of obligations in the same degree. But this is not a statement about women as opposed to men.

The Temporal Mosaic: The Episodic Character of Individual Life

Being individuals slow to reach maturity, we are among the most neotenic of species. This resiliency makes humans prime examples of "K" type species evolution (education, few offspring, slow development). "Culture" constitutes the social contrivances that mitigate neoteny. The transformation of the self through aging is inevitable, but whether we move through successive levels of maturity and the fullest realization of our genome's potential depends on the quality of the active embrace of society in all of the nurturance stages. Incomplete, ontogeny runs to the dead end of immaturity and a miasma of pathological limbos.[100]

The important nurturant occasions are like triggers in epigenesis. Neoteny, the many years of individual immaturity, depends on the hands of society to escape itself. This mitigation of our valuable retardation is in part episodic and social, a matching of the calendars of postnatal embryology by the inventions of caregivers. Occasions make the human adult. If culture in the form of society does not act in the ceremonial, tutoring, and testing response to the personal, epigenetic agenda, we slide into adult infantility—madness. This fantastic arrangement is foreshadowed in the nucleus of every cell. It is an expectancy of the genome, fostered by society, enacted in ecosystems.

Two of the transformative stages of human ontogeny have been studied in detail among living hunter-gatherers—infant-caregiver relationships and adolescent initiation. The archaeological record leaves little doubt that we see in them ancient patterns which may be incompletely addressed in ourselves. Foremost is the bonding-separation dynamic of the first two years. The interaction of infant and mother and infant and other caregivers emerges as a compelling necessity, perhaps the most powerful shaping force in the whole of individual experience. The "social skills" of the newborn and the mother's equally indigenous reciprocity create not only the primary social tie but the paradigm for existential attitudes. The lifelong perception of the world as a "counterplayer"—caring, nourishing, instructing, and protecting, or vindictive, mechanical, and distant—arises here.

The process arises in our earliest experience and is coupled to patterns of response. Hara Marano says, "Newborns come highly equipped for their first intense meetings with their parents, and in particular their mothers. . . . Biologically speaking, today's mothers and babies are two to three million years old. . . . When we put the body of a mother close to her baby, something is turned on that is part of her genetic makeup."[101] Details of the socially embedded rhythms of parenthood vary from culture to culture, but they can hardly improve on the basic style or primary forms found in hunter-gatherer groups. Studies of babies and parents in these societies reveal that the intense early attachment leads not to prolonged dependency but to a better-functioning nervous system and greater success in the separation process.[102]

Something of the same can be said for the whole of ontogeny, especially those passage-markers by which the caregivers celebrate and energize movement across thresholds by the ripe and ready. Notable among these is adolescent initiation, a subject to which a vast body of science and scholarship has been devoted. Yet again it has fundamental forms for which individual psychology is endowed. Much of modern angst has its roots in the modern collapse of this crucial episode in personal development.

Early experience has this formative and episodic quality, with varying degrees of formality in its context. The hunt is one, bringing into play in the individual the most intense emotions and sense of the mysteries of our existence, to be given a catharsis and mediating transformation. The hunt is a pulse of social and personal preparation, addressed to presences unseen, skills and strategies, festive events and religious participation. We cannot become hunter-gatherers as a whole economy, but we can recover the ontogenetic moment. Can five billion people go hunting in a world where these dimensions of human exisistence were played out in a total population of perhaps one million? They can, because the value of the hunt is not in repeated trips but a single leap forward into the heart-structure of the world, the "game" played to rules that reveal ourselves. What is important is to have hunted. It is like having babies; a little of it goes a long way.

Endemic Resources and the Design of a Lifeway—
a Posthistoric Primitivism

In her book *Prehistoric Art in Europe*, N. K. Sandars identifies four strands of the primordial human experience: (1) "The sense of diffused sacredness which may erupt into everyday life," (2) "an order of relationships the categories of which take no account of genetic barriers and which will lead to ideas of metamorphosis inside and outside this life," (3) "unhistorical time," and

(4) "the character or position of the medicine man or shaman."[103]

These are not, of course, removable entities as such, but they constitute aspects of the Paleolithic genius, emergent gestalts from the separate and portable elements of a culture. As ideals not one of these is a regression to obsolescence but a forward step to Heidegger's *dasein*, Merleau-Ponty's and Whitehead's event world, Eliade's centrality of the rites of passage, Odum's redaction of ecological entities as process and relationship. It is not a matter of what ought to be done or how life could be, or even of greater meaning and understanding, but of the nature of experience. I would summarize these "experiences" as follows:

1. Therio-metaphysics. Animals as the language of nature, a great Semiosis. Reading the world as the hunter-gatherer reads tracks. The heuristic principle and hermeneutic act of nature and society as the basic metaphor. Eco-predicated logos.

2. The Voice of Life. Sound, drum, song, voice, instrument, wind, the essential clue to the livingness of the world. It is internal and external at once, the game told as narrative, the play of chance. In story, Snyder has called it "the primacy of together-hearing."

3. The Fledging and Moulting Principle. Epigenesis as the appropriate and sequential coupling of gene and environment, self and other. The ecology of ontogenesis as a resonance between bonding and separation that produces identity. Transitions marked by formal acts of public recognition.

4. Sacramental Trophism. The basic act of communion, transformation, and relatedness, incorporating death as life. It is centered on the act of bringing death and of giving to death as the central celebration of life.

5. The Fire Circle. All forms of social connection in relation to scale. Vernacular gender. Examples: Homeostatic demographic units. The dialectical tribe in Australia: family, band, and tribe affiliation. Sizes 25 to 500. "In terms of conscious dedication to human relationships that are both affective and effective, the primitive is ahead of us all the way," comments Colin Turnbull.[104]

6. Vocational Instruments. Dealing directly with the means of subsistence by hands-on approach. Tools are a gestural response to life, subordinate to thought, art, and religious forms. Marshack speaks of "the demands of fire culture" as one of lore and skills in which the tale is a "metaphysical gift" making the world "an object of contemplation."

7. Place Instead of Space; Moments Replace Time; Chance Instead of

Strategy. Place is at once an external and internal state in a journey home. The place is a process, not coordinates, yet a specific geology, climate, and habitat.

8. Occasions of the Numinous in the relocation of the signs of sacred presence, the mystery of being, and the participatory role of human life, not as ruler or viceroy but as one species of many, in a mood not of guilt or conflict but of affirmation.

9. The escape from domestication, a liberation of nature into itself, including human nature, from the tyranny of the created blobs and the fuzzy goo of emotional—and epoxic glue of ethical—humanism.

Primitivism does not mean a simplified or more thoughtless way of life but a reciprocity with origins, a recovery misconstrued as inaccessible by the ideology of History. In the latter view one puts on costumes and enacts another culture—as the French aristocracy imitated shepherds during the Renaissance or as middle-class "dropouts" in the 1960s put on gingham gowns and bib overalls.

From the ahistoric perspective you cannot "go back" to recover "lost" realities, nor can you completely lose them. So long as there is a green earth and other species our wild genome can make and find its place. Like many difficult things the transformation cannot be made solely by acts of will. One can simulate the external features of a primitive life—for example, the limitation of possessions and the nonownership of the land— but something precedes the outward form and its supporting ideology. That something is the way in which the sensuous apprehension is linked to the conceptual world, the establishment early in life of a mode by which experience and ideas interact, in perception.

It is, of course, a cyclic matter in which childhood experience leads to appropriate thought and custom, which in turn mentors individual genesis. Breaking into the circle is hard, as we urban moderns can only start with an idea of it. Rare are those who can make that leap from the idea to the mode without early shaping. As a result most of us get only glimpses of what we might be were we truer to our wildness, among them some of the anthropologists who study tribal peoples. Or, we get intimations from the archetypes arising in our dreams or given in visionary moments.

In sum, it is an archetypal ecology, a paraprimitive solution, a Paleolithic counterrevolution, a new cynegetics, a venatic mentation. Whatever it may be called, our best guides, when we learn to acknowledge them, will be the living tribal peoples themselves.

NOTES

Acknowledgment: My thanks to Flo Krall for her careful reading, criticism, and suggestions in the preparation of this paper.

1. Herbert J. Muller, *The Uses of the Past* (New York: Oxford University Press, 1952), 38.

2. Herbert Schneidau, *Sacred Discontent* (Los Angeles: University of California, 1976).

3. Bogert O'Brien, "Inuit Ways and the Transformation of Canadian Theology," mss., 1979.

4. Robert Hutchins, Preface to Mortimer J. Adler's Hundred Great Books Series, *The Great Ideas* (Chicago: Encyclopedia Britannica, 1952).

5. Ivan Illich and Barry Sanders, *The Alphabetization of the Popular Mind* (San Francisco: North Point, 1988).

6. Paul Shepard, *Nature and Madness* (San Francisco: Sierra Club Books, 1982).

7. Edmund S. Carpenter, "If Wittgenstein Had Been an Eskimo," *Natural History* 89(4) (February 1980).

8. A. David Napier, *Masks, Transformation, and Paradox* (Los Angeles: University of California, 1986). And see Steven Lonsdale, *Animals and the Origin of Dance* (New York: Thames and Hudson, 1982).

9. Edith Cobb, *The Ecology of Imagination in Childhood* (New York: Columbia University Press, 1978).

10. J. H. Plumb, *Horizon* 41(3) (1972).

11. Paul Shepard, *The Tender Carnivore and the Sacred Game* (New York: Scribners, 1973).

12. George Boas, *Essays on Primitivism and Related Ideas in the Middle Ages* (Baltimore: Johns Hopkins, 1948).

13. Anthony Pagden, *The Fall of Natural Man* (New York: Cambridge University Press, 1982), 78.

14. M. Navarro, *La Tribu Campa* (Lima, 1924), quoted in Gerald Weiss, "Campa Cosmology," *American Museum of Natural History Anthropological Papers* 52 (Part 5) (1975).

15. Will Durant, "A Last Testament to Youth," *The Columbia Dispatch Magazine,* February 8, 1970.

16. Ernest Thompson Seton, *Two Little Savages* (New York: Doubleday, 1903).

17. Calvin Martin, *Keepers of the Game* (Los Angeles: University of California, 1978).

18. Glyn Daniel, *The Idea of Prehistory* (Baltimore: Penguin, 1962), 57.

19. Robert Ardrey, *The Hunting Hypothesis* (New York: Athenaeum, 1976).

20. John Pfeiffer, *The Emergence of Man* (New York: Harper and Row, 1972).

21. Nigel Calder, *Eden Was No Garden* (New York: Holt, 1967).

22. Gordon Rattray Taylor, *Rethink, a Paraprimitive Solution* (New York: Dutton, 1973).

23. José Ortega y Gasset, *Meditations on Hunting* (New York: Scribners, 1972).

24. M. F. C. Bourdillon and Meyer Fortes, eds., *Sacrifice* (New York: Academic Press, 1980).

25. Carleton Coon, *The Story of Man* (New York: Knopf, 1962), 187.

26. W. E. H. Stanner, *White Man Got No Dreaming* (Canberra: Australian National University Press, 1979).

27. Richard B. Lee and Irvin DeVore, eds., *Man the Hunter* (Chicago: Aldine, 1968).

28. Sherwood L. Washburn, ed., *The Social Life of Early Man* (Chicago: Aldine, 1961).

29. G. P. Murdock, *Ethnographic Atlas for New World Societies* (Pittsburgh: University of Pittsburgh Press, 1967).

30. Marshall Sahlins, *Stone Age Economics* (Chicago: Aldine, 1972).

31. Derek Freeman, letter, *Current Anthropology* (October 1973), p. 379.

32. Clifford Geertz, "The Impact of the Concept of Culture on the Concept of Man," in Stanley Diamond, *In Search of the Primitive: A Critique of Civilization* (New Brunswick: Transaction Books, 1974), 102.

33. Melvin Konner, *The Tangled Wing: Biological Constraints on the Human Spirit* (New York: Harper and Row, 1983).

34. Loren Eiseley, "Man of the Future," *The Immense Journey* (New York: Random House, 1957).

35. Laurens van der Post, *Heart of the Hunter* (New York: Harcourt Brace Jovanovich, 1980).

36. Roger M. Keesing, "Paradigms Lost: The New Ethnography and New Linguistics," *South West Journal of Anthropology* 28:299–332 (1972).

37. Ivan Illich, *Gender* (New York: Pantheon, 1982).

38. Gina Bari Kolata, "!Kung Hunter-Gatherers: Feminism, Diet, and Birth Control," *Science* 185:932–934 (1974).

39. Claude Lévi-Strauss, *The Savage Mind* (Chicago: University of Chicago Press, 1966).

40. C. H. D. Clarke, "Venator—the Hunter," mss., n.d.

41. Irenaus Eibes-Eibesfeldt, *Love and Hate* (New York: Holt, 1971).

42. Jane Howard, "All Happy Clans Are Alike," *The Atlantic*, May 1978.

43. Gary Snyder, quoted in Peter B. Chowka, "The Original Mind of Gary Snyder," *East-West*, June 1977.

44. A. H. Ismail and L. B. Trachtman, "Jogging the Imagination," *Psychology Today*, March 1973.

45. P. S. Martin and H. E. Wright, Jr., eds., *Pleistocene Extinctions: The Search for a Cause* (New Haven: Yale University Press, 1967).

46. Donald K. Grayson, "Pleistocene Avifaunas and the Overkill Hypothesis," *Science* 195:691–693 (1977). Karl W. Butzer, *Environment and Archaeology* (Chicago: Aldine, 1971), 503ff. Michael A. Joachim, *Hunter-Gatherer Subsistence and Settlement: A Predictive Model* (New York: Academic Press, 1976). Marvin Harris, "Potlatch Politics and Kings' Castles," *Natural History*, May 1974.

47. Melvin J Konner, "Maternal Care, Infant Behavior and Development Among the !Kung," in R. B. Lee and I. DeVore, *Kalahari Hunter Gatherers* (Cambridge: Harvard University Press, 1976).

48. Joachim, *Hunter-Gatherer Subsistence and Settlement*, 22. Harry Jerison, *Evolution*

of the Brain and Intelligence (New York: Academic, 1973).

49. Richard Nelson, *Make Prayers to the Raven* (Chicago: University of Chicago Press, 1983), 225.

50. Gary Urton, ed., *Animal Myths and Metaphors in South America* (Salt Lake City: University of Utah Press, 1985).

51. Clarke, "Venator—the Hunter."

52. Ibid.

53. James Woodburn, "An Introduction to Haida Ecology," in Lee and DeVore, *Man the Hunter.*

54. Marek Zvelebil, "Postglacial Foraging in the Forests of Europe," *Scientific American,* May 1986.

55. C. Dean Freudenberger, "Agriculture in a Post-Modern World," mss. for conference, "Toward a Post-Modern World," Santa Barbara, California, January 1987.

56. Liberty Hyde Bailey, *The Holy Earth* (New York: Scribner's, 1915), 83.

57. Ibid., 151.

58. Wes Jackson, *New Roots for Agriculture* (San Francisco: Friends of the Earth, 1980).

59. Helen Spurway, "The Causes of Domestication," *Journal of Genetics* 53:325 (1955).

60. Jack M. Potter, *Peasant Society* (Boston: Little Brown, 1967).

61. Jane Schneider, "Of Vigilance and Virgins: Honor, Shame, and Access to Resources in Mediterranean Societies," *Ethnology* 10:1–24 (1971).

62. Aldous Huxley, "Mother," in *Tomorrow and Tomorrow and Tomorrow* (New York: Harper and Row, 1952).

63. Shepard, *Nature and Madness.*

64. Robert Redfield, *The Folk Cultures of Yucatán* (Chicago: University of Chicago Press, 1941).

65. Joseph Campbell, *The Masks of God,* vol. 1 (New York: Viking, 1959), 180.

66. Paul Shepard and Barry Sanders, *The Sacred Paw: The Bear in Nature, Myth and Literature* (New York: Viking, 1985).

67. Tim Ingold, "Hunting, Sacrifice, and the Domestication of Animals," in *The Appropriation of Nature* (Iowa City: University of Iowa Press, 1987).

68. Kolata, "!Kung Hunter-Gatherers."

69. Patricia Draper, "Social and Economic Constraints on Child Life Among the !Kung," in Lee and DeVore, *Kalahari Hunter Gatherers.*

70. C. H. Brown, "Mode of Subsistence and Folk Biological Taxonomy," *Current Anthropology* 26(1): 43–53 (1985).

71. "Campaign to Promote the Vegetarian Diet" (leaflet), Animal Aid Society, Tonbridge, England, n.d.

72. Marvin Harris, *Sacred Cow, Abominable Pig* (New York: Simon and Schuster, 1985), 22.

73. Robert Allen, "Food for Thought," *The Ecologist,* January 1975.

74. Stanley Boyd Eaton and Marjorie Shostak, "Fat Tooth Blues," *Natural History* 95(6) (July 1986).

75. Daniel, *The Idea of Prehistory.*

76. Allen W. Johnson and Timothy Earle, *The Evolution of Human Societies: From Foraging Group to Agrarian State* (Stanford: Stanford University Press, 1987).

77. John W. Berry and Robert C. Annis, "Ecology, Culture and Psychological Differentiation," *International Journal of Psychology* 9:173–193 (1974).

78. Robert Edgerton, *The Individual in Cultural Adaptation* (Los Angeles: University of California Press, 1971).

79. N. K. Sandars, *Prehistoric Art in Europe* (Baltimore: Penguin, 1968), 95–96.

80. Marshall McLuhan, *Through the Vanishing Point* (New York: Harper and Row, 1968).

81. David Lowenthal, "Is Wilderness Paradise Now?," *Columbia University Forum* 7(2) (1964).

82. Susan Sontag, *On Photography* (New York: Dell, 1973).

83. Bertram Lewin, *The Image and the Past* (New York: I.U.P., 1968).

84. Spurway, "The Causes of Domestication."

85. Lévi-Strauss, *The Savage Mind.*

86. Fred Myer, *Pintupi Country, Pintupi Self* (Washington, D.C., Smithsonian Institution, 1986). Myer is following a path laid out by A. Irving Hallowell. For example, see Hallowell, "Self, Society, and Culture in Phylogenetic Perspective," in Sol Tax, ed., *The Evolution of Man* (Chicago: Aldine, 1961).

87. Gary Snyder, "On 'Song of the Taste,'" The Recovery of the Commons Project, Bundle no. 1, North San Juan, Calif., n.d.

88. Gary Snyder, "Good, Wild, Sacred," *Co-Evolution Quarterly*, Fall 1983.

89. René Dubos, "Environmental Determinants of Human Life," in David C. Glass, ed., *Environmental Influences* (Suffern, N.Y.: Rockland University Press, 1968).

90. Ibid., quoted from Lewis Mumford, *The Myth of the Machine* (New York: Harcourt Brace, 1966). Mumford probably got it from Loren Eiseley's "Man of the Future" in *The Immense Journey.*

91. Walter J. Ong, "World as View and World as Event," *American Anthropologist* 71:634–647. Dorothy Lee, "Codifications of Reality: Lineal and Non-Lineal," *Psychosomatic Medicine* 12(2) (1969).

92. Morris Berman, "The Roots of Reality" (a review of Humberto Maturana and Francisco Varelas, *The Tree of Knowledge*), *Journal of Humanistic Psychology* 29:277–284 (1989).

93. Emile Zuckendandle, *Scientific American* 212:63 (1960).

94. Hyemeyohsts Storm, *Storm Arrows* (New York: Harper and Row, 1972).

95. William Ayres Arrowsmith, "Hybris and Sophrosyne," *Dartmouth Alumni Magazine*, July 1970.

96. Wilhelm Dupré, *Religion in Primitive Cultures* (The Hague: Mouton, 1975), 327.

97. Kevin T. Jones, "Hunting and Scavenging by Early Hominids: A Study in Archaeological Method and Theory," Ph.D. dissertation, University of Utah, 1984.

98. Janet Siskind, *To Hunt in the Morning* (New York: Oxford University Press, 1976), 109.

99. Martin King Whyte, *The Status of Women in Preindustrial Societies* (Princeton: Princeton University Press, 1978).

100. Arnold Modell, "The Sense of Identity: The Acceptance of Separateness," in *Object Love and Reality* (New York: I.U.P., 1968).

101. Hara Estroff Marano, "Biology Is One Key to the Bonding of Mothers and Babies," *Smithsonian*, February 1981.

102. Melvin J. Konner, "Maternal Care, Infant Behavior and Development Among the !Kung," in Lee and DeVore, *Kalahari Hunter Gatherers*.

103. Sandars, *Prehistoric Art in Europe*, 26.

104. Colin M. Turnbull, *The Human Cycle* (New York: Simon and Schuster, 1983).

The Wilderness Is
Where My Genome Lives

Genetical biology and medicine are widely heralded as the "physics" of the twenty-first century. We are told that genetic engineering will become the next equivalent of the twentieth-century's subatomic physics. This notion takes its recent thrust from molecular biology and centers on the decoding of the human genome. Progress in mapping genes on the chromosomes is presented in almost daily communiqués from medical research, specifying the location and identity of deleterious genes in both humans and other animals. The prospect being offered by medicine is, of course, that invasive techniques will enable us to replace deleterious genes with preferred alternatives early in the life of the individual, the better genes, grown in bacteria or other organisms, transferred to egg or sperm cells by a tiny vaccination. Attention to this new phase of eugenics is focused on the more sensational aspects of the "war" against disease and heritable disability. It is also expected that the perfect tomato and ultimate cow will become realities.

Pursuit of total health and perfect crops may drive this research machine and its publicity, but at a less conspicuous level what is happening is that the genetic basis and reality of the normal or optimal human individual is being recognized. In the process of decoding the chromosomes, we are finding that what was thought to be "cultural," "environmental," and "learned" has a genetic basis and that education is mostly a kind of facilitation. We are learning finally that being human is heritable rather than attitudinal (witness the failed attempts to make chimpanzees human by rearing them in human homes or foisting speech on them) and that the vaunted human diversity is a large window on rather narrow variations. That all human traits are ultimately genetic (just as all are dependent on appropriate circumstances for their expression) is becoming evident. The century-old debate concerning nature and nurture is not dead, but its formulation is no longer one of alternatives so much as reciprocity.

The human genome is many hundreds of thousands of years old and is "layered," so to speak, like diamonds in clusters of apish pearls which themselves have older genetic settings, antecedents from primate ancestors and others from still more archaic forebears—mammalian, reptilian, ichythian, invertebrate, and bacterial. As a species we are Pleistocene,[1] owing little or nothing to the millennia of urban life, or of rustication with cereals and goats, except perhaps some local shifts in gene frequency associated with resistance to epidemic disease and food allergies, along with a widened flow of genes between physical types or races that were more or less isolated earlier.

While it is clearly demonstratable that genetic change can be extremely rapid in a small, intensely selected population, such as the remnants of a decimated or island group, or among the human-manipulated domestic plants and animals, the evidence of homonid paleontology is that the typical rate of genetic change in human evolution is consistent with that of other wild animals—relatively slow. Whether humans are "domesticated" has been debated for decades, but if we follow the definition of domestic as a type created by controlled breeding with conscious objectives by humans, then we ourselves are genetically wild. The argument that we are domesticated because of our high degree of neoteny or infantilization would require that we regard most species of anthropoid primates as domestic also, and the usefulness of the word begins to drain away.

Domestication typically produces rapid change, hypertrophy, and homozygous recessive traits—which is to say, anomalies, diminished intelligence, and specialized features, at the expense of overall adaptability. It is typical of domesticated forms that they cannot survive except with human care, usually in gardens, farmyards, households, laboratories, or greenhouses.

The implications of this have interesting environmental aspects, since the habitat of domestic animals tends to be architectural habitations combined with other domestic forms—complexes of unstable, disclimaxed landscapes under human dominion. While we, like many other genetically wild animals (such as foxes, crows, and langurs) can live in such places, we and they are not bound by nor necessarily at our best in them. Neither we nor crows are limited to a complex of engineered landscapes or domestic plants and animals to survive, because we and crows have not had our genes muddled by breeders. The radical implication of this is that we, like other wild forms, may actually be less healthy in the domesticated landscapes than in those places to which our DNA remains more closely tuned.

The home of our wildness is both etymologically and biologically wilderness. Although we may define ourselves in terms of culture, language, and so on, it is evident that the context of our being now, as in the past, is wilderness—an environment lacking domestic organisms entirely and to which, one might say, our genes look expectantly for those circumstances which are their optimal ambience. Domesticated forms are inventions, the products of empirical genetic engineering in the past. Immersed though we may be in built and altered surroundings, we are not confined to them, and our human potential may be fulfilled less in such invented landscapes and the behaviors which they entail than in those cultures and places that are shaped more directly by the terms of our evolutionary genesis.

Like crows and foxes, we are omnivorous, edge forms. Unlike them, our swift mobility through places tends to delude our self-appraisal when it comes to obligatory, ecological constraints (at least I presume that foxes and crows are subjectively realistic about limitations). Civilization conceals our innermost need for those complex communities that characterize wilderness, but does not alter that need. Denying the effects of deviating from the world to which we are adapted is part of modern ideology. As René Dubos pointed out more than two decades ago, our adaptability and accommodation to deleterious environments hides our vulnerability to their effects. It is a masking in which we boastfully perceive ourselves as elevated above our progenitors and cousins.

We have become expert at interpreting a wide range of physical and social disorders—everything from war to allergies—as weaknesses (usually temporary) of the social, political, or technological systems, rather than as evidence of ecological dissonance. That we, like bears and cockroaches, can endure deficient environments has been interpreted as evidence of our transcendence of biological specialization, that widely repudiated condition of dinosaurs and all other extinct forms. Social scientists have insisted for three generations that ours is a "generalized" species, while all around us other animals made the mistake of becoming "over specialized." This fable was so patent that, for a century, paleontologists were unable to recognize our ancestors among any of the dozens of hominid-pongid fossils, seeing them all as being irreversibly overdone.[2]

We are not the generalized species we were said to be. This is the same brain and nervous system whose dysfunction now produces epidemic levels of psychopathology in cities. What was a good (and highly specialized) brain for positioning a terrestrial primate in the Pleistocene is evidently maladapted for life in the throes of its own success. Our whole ontogeny or individual development, like the nervous system itself, may be among the most highly specialized biological complexes in existence. The paradox of an apparently unlimited adaptability and extreme specialization will probably untangle its own contradictions in the twenty-first century, as we discover that cultural choices do not exhibit but hide common, underlying physical limitations and requirements.

Such constraints are part of a universal biological heritage, honed to a Pleistocene reality—that is, to the way of life for the three million years commonly said to have ended about ten thousand years ago. In the twentieth century a renewed sense of limitation, necessity of compliance and of human nature, has begun to emerge and to reverse an era of bizarre cultural hallucinations of "no limits" and human domination over or exemption from the "laws of nature." This shift away from the illusion that we can be anything we want to be was foreshadowed in this century by the work of such people as Nico Tinbergen, Konrad Lorenz, Desmond Morris, Lionel Tiger, Robin Fox, S. Boyd Eaton, Marjorie Shostak, Melvin Konner, E. O. Wilson, and Robert Ardrey, all of whom were vilified for "biologizing" the human species and sullying our self-proclaimed superiority to evolution.

* * *

Until recently we have portrayed wilderness in our past in one of two contrasting fictions: the noble savage, such as Adam and Eve and the Greek gods, living in a golden age; or as the Wild Man, a debased figure lurking at the fringes of civilization, destined to consort with the beasts as one of them, an aspect of ourselves which we have repudiated as a grunting caveman. This Wild Man is the grotesque monster of civilized hubris, the embodiment of three thousand years of our fear of the wilderness.

Our "wild" state corresponded to what we wrongly thought to be characteristic of wild animals, mistakenly deduced from watching the demented and stupid beasts of the barnyard. The only hope for human beastliness—our rage, terror, lust, gluttony, and murder—was either religious salvation or a "social contract" that would block all those destructive instincts. This ugly vision of wildness as the dark side of our heritage inclines us to shy away from wilderness as the ground of our being. Wilderness has been valued as the place we test our civilized (i.e., urbanized) selfhood against raw nature, as a landscape aesthetic, as an ethical enclave of biodiversity, or as that refuge in which we hope to have a spiritual experience. But a better idea of people in wild places emerges, in which its practical and sensory terms model the optimum qualities of life in many respects, not only of philosophy, preservation biology, and High Culture, but of food, exercise, and social structure. The time is coming to understand the significance of wilderness, not as an adjunct to the affluent traveler, as leisure for an educated, aesthetic, appreciative class, or as a Noah's ark, but as the psychological and ecological mold of humanity, which continues to be fundamental to our species and ourselves.

Twenty years ago, Hugh Iltis wrote: "Man's love for natural colours, patterns, and harmonies, his preference for forest-grassland ecotones which he recreates wherever he settles, even in drastically different landscapes, must be the result (at least to a very large degree) of Darwinian natural selection through eons of mammalian and anthropoid evolutionary time. . . . Our eyes and ears, noses, brain, and bodies have all been shaped by nature. Would it not then be incredible indeed, if savannas and forest groves, flowers and animals, the multiplicity of environmental components to which our bodies were originally shaped, were not, at the very least, still important to us? Would not such a concept of 'nature' be a major part of what might be called *a basic optimum human environment?*"[3]

This new perspective comes in part from a better understanding of our primate cousins and ancestors, of hunting-gathering societies present and past, and of the conditions of life prior to the first cities and the earliest domestications. Thus have archaeology and anthropology served to revolutionize recent thought. Even so, radical rethinking about *Homo sapiens* from the social sciences is an exception to the twentieth-century mainstream, with its commitment to cultural relativism. As Krober said, "It is differences that we are interested in." And most academic anthropology and its educated public continue to assume that civil-ization is some kind of orthogenic, progressive panacea.

The social bias against species-specific traits and all other naturalisms was a continuation of the dogma of human uniqueness espoused by Western world religions. For centuries the naturalists merely broke their heads against this insular conceit, Darwin's theory of biological evolution being the principal case in point, still "debated" by those who want none of an organic essence, or who want to see it dressed out only as compassion.

* * *

The surprise that this is changing is not because of new respect for the work of naturalists or because official anthropology has turned about, but through the doors of medicine itself. It is all the more surprising because of the essentially unecological stance of modern medicine's adherence to the ideal of the preservation of life at any cost, its official blindness to catastrophic human overpopulation and loss of other species, and its fanatic devotion to antibiotics and expensive technology. Yet, we will be inadvertently convinced of the value of our wildness because our health in the broadest sense depends on it. As we begin to see organic dysfunction and disease as the misfitting of our genome and the environments we have created, we move away from the notion of a "war" against natural process. Acknowledgment that we are indeed Pleistocene hominids, keyed with infinite exactitude to small-group, omnivorous life in semiforested habitats, may not be immediately forthcoming—even from those with the immunological intolerances for milk and cereals, those whose vascular systems are clogged with domestic fats and cholesterols, whose bodies creek with arthritic sedentism, in the midst of epidemic psychoses of overdense populations whose cosmologies yield havoc because they demand control over rather than compliance with the natural world, cosmologies based on the centralized model of the barnyard. We have begun to move toward better diet and exercise because of immediate necessity, but in the long run its measure is life in the lost world of the ice ages, from which such symptoms mark our alienation. The seemingly remote world of the "ice age" and the savannas that preceded it is where the criteria were established that will decide finally whether our medical therapies and "lifestyle" are successful and whether we truly understand what recovery means.

No one, it is said, can go back to the Pleistocene. We will not, in some magic time warp that denies duration, join those prehistoric dead in their well-honed ecology. But that is irrelevant. Having never left our genome and its authority, we have never left the past, which is part of ourselves, and have only to bring the Pleistocene to us. Regardless of the lines drawn to end that period by geologists and archaeologists, we remain "in" it. Fortunately, it is not only a Thing or a Place or a Time but a mosaic lifeway, a living embroidery. The Pleistocene is accessible in its astonishing intimacy and perennial presence. We continue to share the world with most of the families of plants and animals who were also part of it.

A culture is an *assemblage*, not a monolith. The Pleistocene is constructible in terms of its ontogenetic, economic, social, ecological, and cosmological characteristics. The omnivorous mode and small-scale community of human life is not a mono anything—monolatry, monogamy, monopoly, or monotony.

* * *

Perhaps the time has come to dispose of the notion of wilderness as a zoo, an exalted aesthetic, a captive, exotic landscape, or a storehouse of tomorrow's resources. Wildness is the state against which we assess the "virtues" of civilization and its correlates—mass society, the use of fossil fuels, growth-oriented economics, and the technologies of disjunction and pseudomastery that temporarily conceal our limitations and lead us to play in a world of virtual reality rather than live in actual places.

Our hearts are touched by those who seek to create a therapy in the wilderness. But the effort to recreate, to study, or to appreciate the balm of wilderness, to compose a literature of self-discovery and solitude and new awareness is, culturally speaking, merely first a effort, only palliative. Our concern over the increasing rate of extinctions and the worldwide diminishing of biodiversity is, in the end, not altruism, nor ethics, nor charity. Wild species are true Others, the components of wilderness, and at the same time are the external correlates of our inmost selves. Together with their abiotic world, their interactive dynamic is so complex that, when a small part of it—fluid mechanics—was discovered to be vastly complicated, the dismayed physicists cried, "Chaos!" But the naturalists have suspected right all along that the world was not chaotic nor are our brains nearly as complicated as a swamp whose vapors alone throw physicists into a tizzy. As the new genetic mapping inches forward in the next century, the resonance of the two ecologies—the biome and the genome—will be perceived as the way to human health.

NOTES

1. That is, emergents of the past three million years, rather than as the humanists would have it, creations of the Holocene or past ten thousand years.

2. The argument was that none of those heavy-browed, big-jawed, hairy anthropoids could be ancestral because our species could not "evolve" from creatures with all those specialized traits. No relict of *Oreopithecus, Ramapithecus, Australopithecus, Homo habilis,* or *Homo erectus* could be anything but "cousins" of our mysterious forebears. This eternal missing-link mythology vanished, however, when we understood that adaptive neoteny (retarded development) could do just what was required, and—voilà! there in our hands were the bones of our kin.

3. Hugh Iltis, "Flowers and Human Ecology," in Cyril Selmes, ed., *New Movements in the Study and Teaching of Biology* (London: Maurice Temple Smith, 1974)

Acknowledgment of Sources

Grateful acknowledgment is expressed for permission to reprint the following previously published material. The essays appear in their original form, with only minor changes.

"The Ark of the Mind" was originally published in *Parabola, The Magazine of Myth and Tradition*, Summer 1983.

"Animal Rights and Human Rites" was originally published in *The North American Review*, Winter 1974.

"Phyto-Resonance of the True Self" is an unpublished manuscript presented first at "The Healing Dimensions of People-Planet Relations," a symposium at the University of California, Davis, March 24-29, 1994.

"Bears and People" contains excerpts from "Celebrations of the Bear," which first appeared in *The North American Review*, September 1985.

"Searching Out Kindred Spirits" was originally published in *Parabola, The Magazine of Myth and Religion*, May 1991.

"On Animal Friends" appeared in *The Biophilia Hypothesis*, edited by Stephen R. Kellert and Edward O. Wilson. Copyright © 1993 by Island Press.

"The Corvidean Millennium, or A Letter from an Old Crow" was originally published by the University of Chicago Press in *Perspectives in Biology and Medicine*. Copyright © 1964 University of Chicago. Reprinted by permission. All rights reserved.

"Place and Human Development" was published in *Proceedings: Symposium on Children, Nature, and Urban Development*. U.S. Forest Service and the Gifford Pinchot Institute, Washington, D.C., 1977.

"Place in American Culture" was originally published in *The North American Review*, Fall 1977.

Lines from "Ode to Terminus" by W. H. Auden are reprinted with permission from *The New York Review of Books*. Copyright © 1968 Nyrev, Inc.

Lines from a poem by Rebecca West from the introduction to *Carl Sandburg, Selected Poems*, edited by Rebecca West. Published by Harcourt, Brace, 1926. Reprinted by permission.

Lines from "In Praise of Diversity" by Phyllis McGinley are reprinted with permission from *The American Scholar*, Summer 1954. Copyright © 1954 the Phi Beta Kappa Society.

"Ecology and Man—A Viewpoint" by Paul Shepard was originally published in *The Subversive Science*, edited by Paul Shepard and Daniel McKinley. Published by Houghton Mifflin Company. Copyright © 1969 Paul Shepard and Daniel McKinley.

"Advice from the Pleistocene" was originally published as "The Virtues of Anonymity" in *Saturday Review*, September 17, 1966.

"The Philosopher, the Naturalist, and the Agony of the Planet" was originally published in *Carnivore*, Vol. 8, Part 1 (1985). (Good-faith efforts have been made to secure permission to reprint material from this work. If the copyright holder will contact the publisher, any necessary corrections can be made to future printings.)

"Hunting for a Better Ecology" was originally published in *The North American Review*, Summer 1973.

"If You Care About Nature You Can't Go on Hating the Germans Like This" was originally published as "Homage to Heidegger" in *Deep Ecology*, edited by Michael Tobias. Published by Avant Books, 1985. (Good-faith efforts have been made to secure permission to reprint material from this work. If the copyright holder will contact the publisher, any necessary corrections can be made to future printings.)

"Virtually Hunting Reality in the Forests of Simulacra" appeared in *Reinventing Nature*, edited by Michael E. Soulé and Gary Lease. Copyright © 1995 by Island Press.

"A Post-Historic Primitivism" appeared in *The Wilderness Condition*, edited by Max Oelschlaeger. Copyright © 1992 by Max Oelschlaeger. Reprinted by permission of Sierra Club Books.

"The Wilderness is Where My Genome Lives" was published by *Whole Terrain, Reflective Environmental Practice*, Vol. 4, 1995/1996. Copyright © 1995 Antioch New England Graduate School.

Index